T0234573

SpringerBriefs in Optimization

Series Editors

Sergiy Butenko, *Texas A&M University, College Station, TX, USA*
Mirjam Dür, *University of Augsburg, Augsburg, Germany*
Panos M. Pardalos, *University of Florida, Gainesville, FL, USA*
János D. Pintér, *Rutgers University, Piscataway, NJ, USA*
Stephen M. Robinson, *University of Wisconsin-Madison, Madison, WI, USA*
Tamás Terlaky, Lehigh *University, Bethlehem, PA, USA*
My T. Thai ⓘ, *University of Florida, Gainesville, FL, USA*

SpringerBriefs in Optimization showcases algorithmic and theoretical techniques, case studies, and applications within the broad-based field of optimization. Manuscripts related to the ever-growing applications of optimization in applied mathematics, engineering, medicine, economics, and other applied sciences are encouraged.

Alexander J. Zaslavski

Optimization in Banach Spaces

 Springer

Alexander J. Zaslavski
Department of Mathematics
The Technion – Israel Institute of
Technology
Haifa, Israel

ISSN 2190-8354 ISSN 2191-575X (electronic)
SpringerBriefs in Optimization
ISBN 978-3-031-12643-7 ISBN 978-3-031-12644-4 (eBook)
https://doi.org/10.1007/978-3-031-12644-4

Mathematics Subject Classification: 65K10, 90C25, 90C26, 90C48

This Springer imprint is published by the registered company Springer Nature Switzerland AG
The registered company address is: Gewerbestrasse 11, 6330 Cham, Switzerland

Preface

The book is devoted to the study of constrained minimization problems on closed and convex sets in Banach spaces with a Frechet differentiable objective function. Such problems are well studied in a finite-dimensional space and in an infinite-dimensional Hilbert space equipped with an inner product which induces a complete norm. When the space is Hilbert, there are many algorithms for solving optimization problems, including the gradient projection algorithm, which is one of the most important tools in the optimization theory, nonlinear analysis, and their applications.

An optimization problem is described by an objective function and a set of feasible points. For the gradient projection algorithm, each iteration consists of two steps. The first step is a calculation of a gradient of the objective function, while in the second one, we calculate a projection on the feasible set. In each of these two steps, there is a computational error. In general, these two computational errors are different. In our recent research [48, 53, 54], we show that the gradient projection algorithm generates a good approximate solution, if all the computational errors are bounded from above by a small positive constant. Moreover, if we know computational errors for the two steps of the algorithm, we find out what an approximate solution can be obtained and how many iterates one needs for this.

It should be mentioned that in all these works, the properties of a Hilbert space play an important role. When we consider an optimization problem in a general Banach space, the situation becomes more difficult and less understood. On the other hand, such problems arise in the approximation theory.

In this book, our goal is to obtain a good approximate solution of the constrained optimization problem in a general Banach space under the presence of computational errors. It is shown that the algorithm generates a good approximate solution, if the sequence of computational errors is bounded from above by a small constant. The book consists of four chapters. In the first, we discuss several algorithms which are studied in the book and prove a convergence result for an unconstrained problem which is a prototype of our results for the constrained problem. In Chap. 2, we analyze convex optimization problems. Nonconvex optimization problems are studied in Chap. 3. In Chap. 4, we study continuous algorithms for minimization problems under the presence of computational errors.

The author believes that this book will be useful for researchers interested in the optimization theory and its applications.

Haifa, Israel   Alexander J. Zaslavski
February 28, 2022

Contents

1

Introduction

In this book, we study algorithms for constrained minimization problems in a general Banach space. Our goal is to obtain a good approximate solution of the problem in the presence of computational errors. It is shown that the algorithm generates a good approximate solution, if the sequence of computational errors is bounded from above by a small constant. In this section, we discuss several algorithms that are studied in the book. We also prove a convergence result for an unconstrained problem that is a prototype of our results for the constrained problem.

1.1 Notation

In this section, we collect the notation that will be used in the book.

Let $(X, \|\cdot\|)$ be a Banach space equipped with the norm $\|\cdot\|$ that induces the topology in X. We denote by X^* its dual space with the norm $\|\cdot\|_*$. For $x \in X$ and $l \in X^*$, we set $l(x) = \langle l, x \rangle$. The symbol $\langle \cdot, \cdot \rangle$ is referred to as the duality pairing between X^* and X. For each $x \in X$ and each $r > 0$, set

$$B(x, r) = \{y \in X : \rho(x, y) \le r\}.$$

For each function $f : Y \to R^1$, where Y is nonempty, we set

$$\inf(f, Y) = \inf\{f(y) : y \in Y\}.$$

If a real-valued function f is defined in a neighborhood a point x in a Banach space X, then by $f'(x)$ we denote a Frechet derivative of f at x if it exists.

We denote by $\mathrm{mes}(\Omega)$ the Lebesgue measure of a Lebesgue measurable set $\Omega \subset R^1$ and define

$$\chi_\Omega(x) = 1 \text{ for all } x \in \Omega, \ \chi_\Omega(x) = 0 \text{ for all } x \in R^1 \setminus \Omega.$$

© The Author(s), under exclusive license to Springer Nature Switzerland AG 2022
A. J. Zaslavski, *Optimization in Banach Spaces*, SpringerBriefs in Optimization, https://doi.org/10.1007/978-3-031-12644-4_1

In the sequel, we denote by $\mathrm{Card}(D)$ the cardinality of a set D; suppose that the sum over an empty set is zero and that the infimum over an empty set is ∞.

1.2 Constrained Optimization

In our book, we consider a minimization problem

$$f(x) \to \min \qquad (P)$$

$$x \in C,$$

where $C \subset U$ is a convex closed set in a Banach space $(X, \|\cdot\|)$, $f : U \to R^1$ is a convex Frechet differentiable function that is Lipschitz on bounded sets, and U is a convex open set in X. This problem is well studied when the space X is finite-dimensional and when X is an infinite-dimensional Hilbert space equipped with an inner product denoted by $\langle \cdot, \cdot \rangle$ that induces a complete norm $\|\cdot\|$.

When the space X is Hilbert, the problem is solved by using the subgradient projection algorithm that is one of the most important tools in the optimization theory, nonlinear analysis, and their applications. See, for example, [1–3, 5–7, 9–15, 17, 20–26, 28, 31–37, 39–41, 43–47, 49–55]. For this algorithm, we do not need to assume that the function f is Frechet differentiable. Instead of the Frechet derivative, subgradients of a convex function are used [29, 30].

Our problem is described by the objective function f and the set of feasible points C. For the subgradient projection algorithm, each iteration consists of two steps. The first step is a calculation of a subgradient of the objective function, while in the second one, we calculate a projection on the feasible set. In each of these two steps, there is a computational error. In general, these two computational errors are different. In our recent research [48, 53, 54], we show that the subgradient projection algorithm generates a good approximate solution, if all the computational errors are bounded from above by a small positive constant. Moreover, if we know computational errors for the two steps of the algorithm, we find out what an approximate solution can be obtained and how many iterates one needs for this.

It should be mentioned that in all these works, the properties of a Hilbert space play an important role. When X is a general Banach space, the situation becomes more difficult and less understood. Probably the first result for the minimization problem (P) in a Banach space was obtained in [16]. Recently, problem (P) in a Banach space was studied in [18, 19, 42] using greedy algorithms.

Now we describe our algorithm. Let $\lambda \in (0, 1]$ and $c_* \geq 1$ be fixed.

Assume that $x_t \in C$ is a current iteration vector where $t \geq 0$ is an integer. In our first step, we calculate the Frechet derivative $f'(x_t)$. Since we take into account computational errors produced by our computer system instead of $f'(x_t)$, we get its approximation $g_t \in X^*$ satisfying

$$\|g_t - f'(x_t)\| \leq \delta,$$

where $\delta > 0$ is a computational error.

In our second step, we solve the following auxiliary minimization problem:

$$\langle g_t, \eta \rangle \to \min \qquad\qquad (P_a)$$

$$\eta \in B(0, c_*) \cap (C - x_t).$$

Since X is a general Banach space, the existence of a solution of this problem is not guaranteed. Taking into account this fact and also the presence of computational errors at the second step, we get a vector

$$l_t \in C - x$$

such that there exists

$$\xi_t \in B(0, c_*) \cap B(l_t, \delta),$$

which satisfies

$$\langle g_t, \xi_t \rangle \leq \lambda \inf \{ \langle g_t, \eta \rangle : \eta \in B(0, c_*) \cap (C - x_t) \} + \Delta.$$

Here, $\delta, \Delta > 0$ are computational errors produced by our computer system that occur when we solve the auxiliary problem (P_a).

In the third step, we calculate the next iteration $x_{t+1} = x_t + \alpha_t l_t$, where $\alpha_t \in [0, 1]$ is a step size.

In our book, we deal with three options of choosing the step size α_t. In algorithm $(\mathcal{A}1)$, α_t is an approximate solution of the auxiliary optimization problem

$$f(x_t + \alpha l_t) \to \min, \ \alpha \in [0, 1].$$

In algorithm $(\mathcal{A}2)$, it is given a sequence $\{\alpha_t\}_{t=0}^{\infty}$ satisfying

$$\lim_{t \to \infty} \alpha_t = 0 \text{ and } \sum_{t=0}^{\infty} \alpha_t = \infty.$$

In algorithm $(\mathcal{A}3)$, $\alpha_t \in [\beta_1, \beta_0]$, where $0 < \beta_1 < \beta_0$ are given constants. (It is possible that $\beta_1 = \beta_0$, of course.)

Since we take into account the computation errors, the inequality $f(x_t + \alpha_t l_t) \leq f(x_t)$ cannot be guaranteed. As a result, each of algorithms $(\mathcal{A}1)$–$(\mathcal{A}3)$ has two subcases: in the first case, we always define $x_{t+1} = x_t + \alpha_t l_t$, while in the second subcase, we define x_{t+1} by the equality above only if

$$f(x_t + \alpha_t l_t) \leq f(x_t);$$

otherwise, $x_{t+1} = x_t$. So actually we have six different algorithms.

The discussion above leads us to the following definition that allows us to describe the second steps of our algorithms shortly.

For each $x \in C$ and each pair of numbers $\delta, \Delta \in (0, 1]$, denote by $E(x, \delta, \Delta)$ the set of all $l \in C - x$ for which there exist $g \in X^*$ such that

$$\|g - f'(x)\|_* \leq \delta$$

and

$$\xi \in B(0, c_*) \cap B(l, \delta)$$

such that

$$\langle g, \xi \rangle \leq \lambda \inf\{\langle g, \eta \rangle : \eta \in B(0, c_*) \cap (C - x)\} + \Delta.$$

With this definition, the second step of our algorithms is described as the choice of

$$l_t \in E(x_t, \delta, \Delta).$$

Section 2.3 contains Theorems 2.2–2.10 that show the behavior of the algorithms. In these results, we show that for a given $\epsilon > 0$, there exist a sufficiently small error $\delta, \Delta > 0$ such that after a certain number of iterations we obtain a point $x_s \in K$ satisfying $f(x_s) \leq \inf(f, C) + \epsilon$. Of course, it is interesting to obtain an explicit estimation for δ, Δ. It is done in Theorem 2.11 (see Sect. 2.13) under some additional assumptions on the objective function f. Namely, we assume that its Frechet derivative is Holder continuous on bounded sets and that the corresponding constants are known. In this case, we obtain an explicit dependence δ, Δ and a number of iterations on ϵ. This dependence allows us easily to solve an inverse problem: if we know δ, Δ what ϵ can be obtained?

Note that in [16] it was considered an exact version of algorithm $\mathcal{A}1$ when the set C is bounded. Implicitly, it was assumed that the functional $f'(x_t)$ has a minimizer on C. This is true when C is weakly compact. In [18, 42], also algorithm $\mathcal{A}1$ is studied. In both of these works [18, 42] at the second step, the direction

$$l_t \in C - x$$

satisfies

$$\langle f'(x_t), l_t \rangle \leq \lambda \inf\{\langle f'(x_t), \eta \rangle : \eta \in C - x_t\}.$$

In [42], the step size α_t is an exact solution on the auxiliary minimization problem, while in [18], it is an approximate one.

In the next section of this chapter, we prove a convergence result in the case of unconstrained problems ($C = X$) that is a prototype of our results for constrained problems. It should be mentioned that unconstrained problems were studied in Chapter 8 of [38] using algorithms induced by regular vector fields.

In Chap. 3 of the book, we study nonconvex minimization problems. Continuous versions of our algorithms are studied in Chap. 4.

1.3 Unconstrained Optimization

Assume that $(X, \| \cdot \|)$ is a Banach space and that $f : X \to R^1$ is a convex function that is Lipschitz on all bounded sets in X. We assume that f is Frechet differentiable at any point $x \in X$ such that $f(x) > \inf(f, X)$. If $x \in X$ satisfies $f(x) = \inf(f, X)$, then we set $f'(x) = 0$.

We assume that the following two assumptions hold:

(A1) There exists a nonempty bounded set $X_0 \subset X$ such that $\inf(f, X_0) = \inf(f, X)$.

(A2) For every pair $M, \epsilon > 0$, the Frechet derivative $f'(\cdot)$ is uniformly continuous on the set

$$\{x \in B(0, M) : f(x) > \inf(f, X) + \epsilon\}.$$

In this section, we consider the minimization problem

$$f(x) \to \min, \ x \in X.$$

In order to solve this problem, we use two iterative processes. Let $c_* > 1$, $\lambda \in (0, 1]$ and $\{\alpha_t\}_{t=0}^{\infty}$, $\{\epsilon_t\}_{t=0}^{\infty} \subset (0, 1]$ satisfy

$$\lim_{t \to \infty} \epsilon_t = 0, \ \lim_{t \to \infty} \alpha_t = 0, \ \sum_{t=0}^{\infty} \alpha_t = \infty.$$

In both processes, we select an arbitrary $x_0 \in X$. Given a current iteration vector $x_t \in X$, we calculate $f'(x_t)$. If $f'(x_t) = 0$, then x_t is a solution of our minimization problem. Assume that $f'(x_t) \neq 0$. We find $l_t \in X$ satisfying

$$\|l_t\| \leq c_*$$

and

$$\langle f'(x_t), l_t \rangle \leq -\lambda \|f'(x_t)\|_* + \epsilon_t.$$

In the first process, we find $x_{t+1} \in \{x_t + \alpha l_t : \alpha \in [0, 1]\}$ such that

$$f(x_{t+1}) \leq f(x_t + \alpha l_t), \ \alpha \in [0, 1].$$

In the second process, if $f(x_t + \alpha_t l_t) < f(x_t)$, then we set $x_{t+1} = x_t + \alpha_t l_t$; otherwise $x_{t+1} = x_t$.

In Sect. 1.5, we prove the following result.

Theorem 1.1. *Let $c_* > 1$, $\lambda \in (0, 1]$ and $\{\epsilon_t\}_{t=0}^{\infty} \subset (0, 1]$ satisfy*

$$\lim_{\to \infty} \epsilon_t = 0. \tag{1.1}$$

Assume that $\{x_t\}_{t=0}^{\infty} \subset C$, $\{l_t\}_{t=0}^{\infty} \subset X$ and that for each integer $t \geq 0$,

$$\|l_t\| \leq c_*, \ \langle f'(x_t), l_t \rangle \leq -\lambda \|f'(x_t)\|_* + \epsilon_t. \tag{1.2}$$

Then the following assertions hold:

1. If for each integer $t \geq 0$,

$$x_{t+1} \in \{x_t + \alpha l_t : \alpha \in [0,1]\}$$

and

$$f(x_{t+1}) \leq f(x_t + \alpha l_t), \ \alpha \in [0,1]$$

and if $\liminf_{t \to \infty} \|x_t\| < \infty$, then

$$\lim_{t \to \infty} f(x_t) = \inf(f, X).$$

2. Assume that $\{\alpha_t\}_{t=0}^{\infty} \subset (0,1]$ satisfies

$$\lim_{\to \infty} \alpha_t = 0, \ \sum_{t=0}^{\infty} \alpha_t = \infty, \tag{1.3}$$

that the sequence $\{x_t\}_{t=0}^{\infty}$ is bounded, and that for each integer $t \geq 0$, if $f(x_t + \alpha_t l_t) < f(x_t)$, then $x_{t+1} = x_t + \alpha_t l_t$; otherwise, $x_{t+1} = x_t$. Then $\lim_{t \to \infty} f(x_t) = \inf(f, X).$

(Note that if $\lim_{\|x\| \to \infty} f(x) = \infty$, then the sequence $\{x_t\}_{t=0}^{\infty}$ is bounded.)

The proof of Theorem 1.1 is based on Lemma 1.2 that proved in the next section.

1.4 An Auxiliary Result

Lemma 1.2. *Let $\lambda \in (0,1]$, $c_* > 1$, $\bar{K}, \bar{\epsilon} > 0$, and*

$$\inf(f, B(0, \bar{K})) = \inf(f, X). \tag{1.4}$$

Then there exist $\bar{\alpha}, \gamma, \Delta \in (0,1)$ such that for each $x \in B(0, \bar{K})$ satisfying

$$f(x) \geq \inf(f, X) + \bar{\epsilon}, \tag{1.5}$$

each $l \in X$ satisfying

$$\|l\| \leq c_*, \ \langle f'(x), l \rangle \leq -\lambda \|f'(x)\|_* + \Delta, \tag{1.6}$$

and each $\beta \in (0, \bar{\alpha}]$, the inequality $f(x) - f(x + \beta l) \geq \beta \gamma$ holds.

Proof. We may assume without loss of generality that $\bar{\epsilon} < 1 < \bar{K}$. Since f is Lipschitz on bounded sets, there exists $L_0 > 1$ such that

$$|f(z_1) - f(z_2)| \leq L_0 \|z_1 - z_2\|, \ z_1, z_2 \in B(0, \bar{K} + 1). \tag{1.7}$$

Choose positive numbers

$$\Delta \leq (16\bar{K})^{-1}\bar{\epsilon}\lambda \tag{1.8}$$

and

$$\gamma \leq (8\bar{K})^{-1}\bar{\epsilon}\lambda. \tag{1.9}$$

By assumption (A2), there exists

$$\bar{\alpha} \in (0, (2L_0 + 2)^{-1}\bar{\epsilon}c_*^{-1}) \tag{1.10}$$

such that the following property holds:

(i) For each $z_1, z_2 \in B(0, \bar{K} + 1)$ satisfying

$$f(z_i) \geq \inf(f, X) + \bar{\epsilon}/4, \ i = 1, 2$$

and $\|z_1 - z_2\| \leq \bar{\alpha}L_0 c_*$, we have

$$\|f'(z_1) - f'(z_2)\|_* \leq (4c_*)^{-1}\Delta.$$

Assume that

$$x \in B(0, \bar{K}) \tag{1.11}$$

satisfies (1.5), $l \in X$ satisfies (1.6), and that

$$\beta \in (0, \bar{\alpha}]. \tag{1.12}$$

In view of (1.5), $f'(x) \neq 0$. By (1.4), there exists

$$x_0 \in B(0, \bar{K}) \tag{1.13}$$

such that

$$f(x_0) \leq \inf(f; X) + 2^{-1}\bar{\epsilon}. \tag{1.14}$$

It follows from (1.5), (1.11), (1.13), (1.14) and the convexity of f that

$$-\bar{\epsilon}/2 \geq f(x_0) - f(x) \geq \langle f'(x), x - x_0 \rangle \geq -2\bar{K}\|f'(x)\|_*$$

and

$$\|f'(x)\|_* \geq (4\bar{K})^{-1}\bar{\epsilon}. \tag{1.15}$$

We estimate

$$f(x + \beta l) - f(x).$$

By the mean value theorem, there exists $\theta \in [0, \beta]$ such that

$$f(x + \beta l) - f(x) = \langle f'(x + \theta l), \beta l \rangle. \tag{1.16}$$

Equations (1.6) and (1.16) imply that

$$f(x + \beta l) - f(x) \leq \langle \beta f'(x), l \rangle + \|f'(x + \theta l) - f'(x)\|_* \beta \|l\|$$

$$\leq \beta(-\lambda \|f'(x)\|_* + \Delta) + \beta \|f'(x + \theta l) - f'(x)\|_* c_*. \qquad (1.17)$$

It follows from (1.6) and (1.10)–(1.12) that

$$\|x + \theta l\| \leq \|x\| + \|l\|\bar{\alpha} \leq \bar{K} + 1. \qquad (1.18)$$

In view of (1.6), (1.7), (1.11), (1.12), and (1.18),

$$|f(x + \theta l) - f(x)| \leq L_0 \beta \|l\| \leq \bar{\alpha} L_0 c_*. \qquad (1.19)$$

By (1.5), (1.10), and (1.19),

$$f(x + \theta l) \geq f(x) - \bar{\alpha} L_0 c_*$$

$$\geq \inf(f, X) + \bar{\epsilon} - \bar{\alpha} L_0 c_* \geq \inf(f, X) + \bar{\epsilon}/2. \qquad (1.20)$$

Property (i) and equations (1.5), (1.6), (1.11), (1.12), (1.18), and (1.20) imply that

$$\|f'(x + \theta l) - f'(x)\|_* \leq (4c_*)^{-1}\Delta. \qquad (1.21)$$

By (1.10), (1.12), (1.15), (1.17)–(1.19), and (1.21),

$$f(x + \beta l) - f(x) \leq \beta(-\lambda \|f'(x)\|_* + \Delta) + \beta \Delta/4$$

$$\leq \beta(2\Delta - \lambda \|f'(x)\|_*) \leq \beta(2\Delta - (4\bar{K})^{-1}\bar{\epsilon}\lambda)$$

$$\leq -(8\bar{K})^{-1}\bar{\epsilon}\lambda\beta \leq -\gamma\beta$$

and

$$f(x + \beta l) \leq f(x) - \beta\gamma.$$

Lemma 1.2 is proved.

1.5 Proof of Theorem 1.1

Let us prove assertion 1. Assume that for each integer $t \geq 0$,

$$x_{t+1} \in \{x_t + \alpha l_t : \ \alpha \in [0, 1]\}, \qquad (1.22)$$

$$f(x_{t+1}) \leq f(x_t + \alpha l_t), \ \alpha \in [0, 1], \qquad (1.23)$$

and

$$\liminf_{t\to\infty} \|x_t\| < \infty. \tag{1.24}$$

We show that

$$\lim_{t\to\infty} f(x_t) = \inf(f, X).$$

Assume the contrary. Then there exists $\epsilon > 0$ such that

$$f(x_t) \geq \inf(f, C) + \epsilon, \ t = 0, 1, \ldots \tag{1.25}$$

In view of (1.24), there exist $\bar{K} > 0$ and a subsequence $\{x_{i_k}\}_{k=1}^\infty$ such that

$$\|x_{i_k}\| \leq \bar{K}, \ k = 1, 2, \ldots \tag{1.26}$$

In view of (A1), we may assume without loss of generality that

$$\inf(f, B(0, \bar{K})) = \inf(f, X). \tag{1.27}$$

Lemma 1.2 implies that there exist $\bar{\alpha}, \gamma, \Delta \in (0, 1)$ such that the following property holds:

(i) For each $x \in B(0, \bar{K})$ satisfying

$$f(x) \geq \inf(f, X) + \epsilon,$$

each $l \in X$ satisfying

$$\|l\| \leq c_*, \ \langle f'(x), l \rangle \leq -\lambda \|f'(x)\|_* + \Delta,$$

and each $\beta \in (0, \bar{\alpha}]$, the inequality

$$f(x) - f(x + \beta l) \geq \beta \gamma$$

holds.

In view of (1.1), we may assume without loss of generality that for each integer $k \geq 1$,

$$\epsilon_{i_k} < \Delta. \tag{1.28}$$

Let $k \geq 1$ be an integer. Property (i) and Eqs. (1.2), (1.25), (1.26), and (1.28) imply that

$$f(x_{i_k}) - f(x_{i_k} + \bar{\alpha} l_{i_k}) \geq \bar{\alpha} \gamma. \tag{1.29}$$

It follows from (1.22), (1.23), and (1.29) that for each integer $n \geq 2$,

$$f(x_0) - \inf(f, X) \geq f(x_{i_1}) - f(x_{i_n})$$

$$= \sum_{i=i_1}^{i_n-1} (f(x_i) - f(x_{i+1})) \geq \sum_{k=1}^{n-1} (f(x_{i_k}) - f(x_{i_k+1}))$$

$$\geq \sum_{k=1}^{n-1}(f(x_{i_k}) - f(x_{i_k} + \bar{\alpha}l_{i_k})) \geq \bar{\alpha}\gamma n \to \infty \text{ as } n \to \infty.$$

The contradiction we have reached completes the proof of assertion 1. Let us prove assertion 2. Assume that $\{\alpha_t\}_{t=0}^{\infty} \subset (0,1]$ satisfies

$$\lim_{t\to\infty} \alpha_t = 0, \ \sum_{t=0}^{\infty} \alpha_t = \infty, \tag{1.30}$$

the sequence $\{x_t\}_{t=0}^{\infty}$ is bounded and that for each integer $t \geq 0$,

if $f(x_t + \alpha_t l_t) < f(x_t)$, then $x_{t+1} = x_t + \alpha_t l_t$; otherwise, $x_{t+1} = x_t$.
$$\tag{1.31}$$

In order to complete the proof, it is sufficient to show that

$$\lim_{t\to\infty} f(x_t) = \inf(f, X).$$

Assume the contrary. Then by (1.31), there exists $\epsilon > 0$ such that

$$f(x_t) > \inf(f, X) + \epsilon, \ t = 0, 1, \ldots \tag{1.32}$$

There exists $\bar{K} > 0$ such that

$$\|x_i\| \leq \bar{K}, \ i = 1, 2, \ldots \tag{1.33}$$

In view of (A1), we may assume without loss of generality that

$$\inf(f, B(0, \bar{K})) = \inf(f, X). \tag{1.34}$$

Lemmas 1.2 and (1.34) imply that there exist $\bar{\alpha}, \gamma, \Delta \in (0,1)$ such that the following property holds:
(ii) For each $x \in B(0, \bar{K})$ satisfying

$$f(x) \geq \inf(f, X) + \epsilon,$$

each $l \in X$ satisfying

$$\|l\| \leq c_*, \ \langle f'(x), l \rangle \leq -\lambda\|f'(x)\|_* + \Delta,$$

and each $\beta \in (0, \bar{\alpha}]$, the inequality

$$f(x) - f(x + \beta l) \geq \beta\gamma$$

holds.

In view of (1.1) and (1.30), we may assume without loss of generality that for each integer $t \geq 0$,

$$\epsilon_t \leq \Delta, \ \alpha_t \leq \bar{\alpha}. \tag{1.35}$$

Property (ii) and Eqs. (1.32), (1.33), and (1.35) imply that for each integer $t \geq 0$,

$$f(x_t) - f(x_t + \alpha_t l_t) \geq \alpha_t \gamma. \tag{1.36}$$

It follows from (1.31) and (1.36) that for each integer $t \geq 0$,

$$x_{t+1} = x_t + \alpha_t l_t$$

and

$$f(x_t) - f(x_{t+1}) \geq \alpha_t \gamma. \tag{1.37}$$

By (1.30) and (1.37), for each integer $n \geq 1$,

$$f(x_0) - \inf(f, X) \geq f(x_0) - f(x_n)$$

$$= \sum_{i=0}^{n-1} (f(x_i) - f(x_{i+1})) \geq \gamma \sum_{i=0}^{n-1} \alpha_i \to \infty \text{ as } n \to \infty.$$

The contradiction we have reached completes the proof of assertion 2 and Theorem 1.1.

2

Convex Optimization

In this chapter, we study the algorithms for constrained convex minimization problems in a general Banach space that were described in Chap. 1. Our goal is to obtain a good approximate solution of the problem in the presence of computational errors. It is shown that the algorithm generates a good approximate solution, if the sequence of computational errors is bounded from above by a small constant. We obtain a number of convergence results under different conditions including a theorem with explicit estimations for computational errors.

2.1 Preliminaries

Assume that $(X, \|\cdot\|)$ is a Banach space, $U \subset X$ is an open convex set,

$$C \subset U$$

is a nonempty closed and convex set, and $f : U \to R^1$ is a convex function that is Lipschitz on all bounded sets in U. We assume that f is Frechet differentiable at any point $x \in U$ such that $f(x) > \inf(f, C)$.

We assume that the following two assumptions hold:

(A1) There exists a nonempty bounded set $C_0 \subset C$ such that $\inf(f, C_0) = \inf(f, C)$.

(A2) For every pair $M, \epsilon > 0$, the Frechet derivative $f'(\cdot)$ is uniformly continuous on the set

$$\{x \in B(0, M) \cap C : f(x) > \inf(f, C) + \epsilon\}.$$

© The Author(s), under exclusive license to Springer Nature Switzerland AG 2022

A. J. Zaslavski, *Optimization in Banach Spaces*, SpringerBriefs in Optimization, https://doi.org/10.1007/978-3-031-12644-4_2

In this section, we consider the minimization problem

$$f(x) \to \min, \ x \in C. \tag{P}$$

Let $\lambda \in (0, 1]$ and $c_* \geq 1$ be fixed. For each $x \in C$ and each pair of numbers $\delta, \Delta \in (0, 1]$, denote by $E(x, \delta, \Delta)$ the set of all $l \in C - x$ for which there exist $g \in X^*$ such that

$$\|g - f'(x)\|_* \leq \delta \tag{2.1}$$

and

$$\xi \in B(0, c_*) \cap B(l, \delta)$$

such that

$$\langle g, \xi \rangle \leq \lambda \inf\{\langle g, \eta \rangle : \eta \in B(0, c_*) \cap (C - x)\} + \Delta. \tag{2.2}$$

For problem (P), we use algorithms $\mathcal{A}1 - \mathcal{A}3$. Their analysis is based on a lemma that is proved in the next section.

2.2 A basic Lemma

Lemma 2.1. *Let* $\lambda \in (0, 1]$, $c_* > 1$, $\bar{K} \geq 1$, $\bar{\epsilon} \in (0, 1]$, *and*

$$\inf(f, C \cap B(0, \bar{K})) = \inf(f, C). \tag{2.3}$$

Then there exist $\bar{\alpha}, \gamma, \Delta, \delta \in (0, 1)$ *such that for each* $x \in B(0, \bar{K}) \cap C$ *satisfying*

$$f(x) \geq \inf(f, C) + \bar{\epsilon},$$

each $l \in E(x, \delta, \Delta)$ *and each* $\beta \in (0, \bar{\alpha}]$, *the inequality*

$$f(x) - f(x + \beta l) \geq \beta \gamma$$

holds.

Proof. Since f is Lipschitz on bounded sets, there exists $L_0 \geq 1$ such that

$$|f(z_1) - f(z_2)| \leq L_0 \|z_1 - z_2\|, \ z_1, z_2 \in B(0, \bar{K} + 1) \cap U. \tag{2.4}$$

Fix a positive number

$$\delta_1 \leq (32\bar{K}c_*)^{-1}\bar{\epsilon}\lambda. \tag{2.5}$$

By assumption (A2), there exists

$$\delta_2 \in (0, \delta_1)$$

such that the following property holds:

(a) For each pair of points $z_1, z_2 \in B(0, \bar{K} + 1) \cap C$ satisfying

$$f(z_i) \geq \inf(f, C) + \bar{\epsilon}/4, \; i = 1, 2$$

and

$$\|z_1 - z_2\| \leq \delta_2,$$

we have

$$\|f'(z_1) - f'(z_2)\|_* \leq \delta_1.$$

Choose positive numbers

$$\delta \leq (16\bar{K})^{-1} \bar{\epsilon} \lambda (L_0 + 3c_*)^{-1}, \tag{2.6}$$

$$\Delta < (16\bar{K})^{-1} \bar{\epsilon} \lambda, \tag{2.7}$$

$$\gamma \leq (16\bar{K})^{-1} \bar{\epsilon} \lambda, \tag{2.8}$$

and

$$\bar{\alpha} \leq \min\{(2c_*)^{-1} \delta_2, \; (4c_* L_0)^{-1} \bar{\epsilon}\}. \tag{2.9}$$

Assume that

$$x \in B(0, \bar{K}) \cap C \tag{2.10}$$

satisfies

$$f(x) \geq \inf(f, C) + \bar{\epsilon}, \tag{2.11}$$

$$l \in E(x, \delta, \Delta), \tag{2.12}$$

and

$$\beta \in (0, \bar{\alpha}]. \tag{2.13}$$

By (2.3), there exists

$$x_0 \in B(0, \bar{K}) \cap C \tag{2.14}$$

such that

$$f(x_0) \leq \inf(f, C) + 2^{-1} \bar{\epsilon}. \tag{2.15}$$

It follows from (2.11), (2.15) and the convexity of f that

$$-\bar{\epsilon}/2 \geq f(x_0) - f(x) \geq \langle f'(x), x - x_0 \rangle. \tag{2.16}$$

In view of (2.10) and (2.14),

$$\|x - x_0\| \leq 2\bar{K}. \tag{2.17}$$

Set

$$x_1 = x + (2\bar{K})^{-1}(x_0 - x) = (2\bar{K})^{-1} x_0 + (1 - (2\bar{K})^{-1}) x \in C \tag{2.18}$$

and set

$$\eta = x_1 - x. \tag{2.19}$$

It follows from (2.17)–(2.19) that

$$\eta \in C - x, \ \|\eta\| \leq 1. \tag{2.20}$$

By (2.16), (2.18), (2.19) and convexity of f,

$$\langle f'(x), \eta \rangle = \langle f'(x), x_1 - x \rangle \leq f(x_1) - f(x)$$

$$\leq (2\bar{K})^{-1} f(x_0) + (1 - (2\bar{K})^{-1}) f(x) - f(x) = (2\bar{K})^{-1} (f(x_0) - f(x)) \leq -(4\bar{K})\bar{\epsilon}. \tag{2.21}$$

It follows from (2.1), (2.2), (2.12), and (2.20) that there exist

$$\xi \in B(0, c_*) \cap B(l, \delta) \tag{2.22}$$

and $g \in X^*$ such that

$$\|g - f'(x)\|_* \leq \delta \tag{2.23}$$

and

$$\langle g, \xi \rangle \leq \lambda \langle g, \eta \rangle + \Delta. \tag{2.24}$$

By the mean value theorem, there exists $\theta \in [0, \beta]$ such that

$$f(x + \beta l) - f(x) = \langle f'(x + \theta l), \beta l \rangle. \tag{2.25}$$

Equations (2.9), (2.12), and (2.13) imply that

$$0 \leq \beta \leq \bar{\alpha} < 1, l \in C - x$$

and

$$x + \theta l, \ x + \beta l \in C. \tag{2.26}$$

In view of (2.6) and (2.22),

$$\|l\| \leq \|\xi\| + 1 \leq c_* + 1 \leq 2c_*. \tag{2.27}$$

By (2.9), (2.13), and (2.27),

$$\|\theta l\| \leq \|\beta l\| \leq 2c_* \bar{\alpha} \leq \delta_2. \tag{2.28}$$

It follows from (2.9), (2.10), (2.13), and (2.27) that

$$\|x + \theta l\| \leq \|x\| + \bar{\alpha}\|l\| \leq \bar{K} + 2c_* \bar{\alpha} \leq \bar{K} + 1. \tag{2.29}$$

Equations (2.4), (2.16), (2.26), (2.28), and (2.29) imply that

$$|f(x + \theta l) - f(x)| \leq L_0 \|\beta l\| \leq 2\bar{\alpha} L_0 c_*. \tag{2.30}$$

By (2.9), (2.11), and (2.30),

$$f(x + \theta l) \geq f(x) - 2\bar{\alpha}L_0 c_*$$

$$\geq \inf(f, C) + \bar{\epsilon} - 2\bar{\alpha}L_0 c_* \geq \inf(f, C) + \bar{\epsilon}/2. \tag{2.31}$$

In view of (2.25),

$$f(x + \beta l) - f(x) \leq \beta\langle f'(x), l\rangle + \|f'(x + \theta l) - f'(x)\|_* \beta\|l\|. \tag{2.32}$$

Property (a) applied with $z_1 = x$, $z_2 = x + \theta l$ and Eqs. (2.10), (2.11), (2.26), (2.28), (2.29), and (2.31) imply that

$$\|f'(x + \theta l) - f'(x)\|_* \leq \delta_1. \tag{2.33}$$

Equations (2.27), (2.32), and (2.33) imply that

$$f(x + \beta l) - f(x) \leq \beta\langle f'(x), l\rangle + 2c_*\beta\delta_1. \tag{2.34}$$

It follows from (2.4), (2.10), and (2.23) that

$$\beta\langle f'(x), l\rangle \leq \beta\langle f'(x), \xi\rangle + \beta\|f'(x)\|_*\|l - \xi\|$$

$$\leq \beta\langle f'(x), \xi\rangle + \beta\delta L_0. \tag{2.35}$$

By (2.20)–(2.24),

$$\langle f'(x), \xi\rangle \leq \langle g, \xi\rangle + \|f'(x) - g\|_*\|\xi\| \leq \langle g, \xi\rangle + + c_*\delta$$

$$\leq \lambda\langle g, \eta\rangle + \Delta + \delta c_*$$

$$\leq \lambda\langle f'(x), \eta\rangle + \|f'(x) - g\|_*\|\eta\| + \Delta + \delta c_*$$

$$\leq -\lambda(4\bar{K})^{-1}\bar{\epsilon} + \Delta + \delta(1 + c_*). \tag{2.36}$$

It follows from (2.5)–(2.8) and (2.34)–(2.36) that

$$f(x + \beta l) - f(x)$$

$$\leq 2\beta c_*\delta_1 + \beta\delta L_0 + \beta[-\lambda(4\bar{K})^{-1}\bar{\epsilon} + 2\delta c_* + \Delta]$$

$$= \beta(-\lambda(4\bar{K})^{-1}\bar{\epsilon} + 2\delta_1 c_* + \delta(L_0 + 2c_*) + \Delta)$$

$$\leq -\beta\lambda(16\bar{K})^{-1}\bar{\epsilon} \leq -\beta\gamma.$$

Lemma 2.1 is proved.

2.3 Convergence Results

We continue to use the notation, definitions, and assumptions introduced in Sect. 2.1. We assume that $c_* \geq 1$ and $\lambda \in (0, 1]$ are fixed. In this section, we state our results that describe the behavior of algorithms $\mathcal{A}1 - \mathcal{A}3$. In our first result related to algorithm $\mathcal{A}1$, we assume that the objective function f satisfies a growth condition. It is proved in Sect. 2.4.

Theorem 2.2. *Assume that*

$$\lim_{\|x\| \to \infty, \, x \in U} f(x) = \infty,$$

and $K, \epsilon > 0$. Then there exist a natural number N_0 and $\delta, \Delta \in (0, 1)$ such that for each sequence $\{x_i\}_{i=0}^{\infty} \subset C$ and each sequence $\{\xi_i\}_{i=0}^{\infty} \subset X$ that satisfy

$$\|x_0\| \leq K$$

and for each integer $i \geq 0$,

$$\xi_i \in E(x_i, \delta, \Delta),$$

$$x_{i+1} \in \{x_i + \alpha \xi_i : \ \alpha \in [0, 1]\},$$

and

$$f(x_{i+1}) \leq f(x_i + \alpha \xi_i) + \delta, \ \alpha \in [0, 1],$$

the inequality

$$f(x_i) \leq \inf(f, C) + \epsilon$$

holds for all integers $i \geq N_0$.

The next result related to algorithm $\mathcal{A}2$ is proved in Sect. 2.5. We assumed that the objective function f satisfies the growth condition and define $x_{t+1} = x_t + \alpha_t \xi_t$ only if $f(x_t + \alpha_t \xi_t) < f(x_t)$; otherwise, $x_{t+1} = x_t$.

Theorem 2.3. *Assume that*

$$\lim_{\|x\| \to \infty, \, x \in U} f(x) = \infty,$$

$\{\alpha_t\}_{t=0}^{\infty} \subset (0, 1]$ *satisfies*

$$\lim_{t \to \infty} \alpha_t = 0, \ \sum_{t=0}^{\infty} \alpha_t = \infty,$$

and $K, \epsilon > 0$. Then there exist a natural number N_0 and $\delta, \Delta \in (0, 1)$ such that for each sequence $\{x_i\}_{i=0}^{\infty} \subset C$ and each sequence $\{\xi_i\}_{i=0}^{\infty} \subset X$ that satisfy

$$\|x_0\| \leq K$$

and for each integer $i \geq 0$,

$$\xi_i \in E(x_i, \delta, \Delta)$$

and

$$x_{i+1} = x_i + \alpha_i \xi_i \ if \ f(x_i + \alpha_i \xi_i) < f(x_i), \ \ otherwise \ x_{i+1} = x_i,$$

the inequality

$$f(x_i) \leq \inf(f, C) + \epsilon$$

holds for all integers $i \geq N_0$.

In the following result related to algorithm $\mathcal{A}1$, the sequence of iterates $\{x_t\}_{t=0}^{\infty}$ is assumed to be bounded. It is proved in Sect. 2.6.

Theorem 2.4. *Assume that $\{\delta_i\}_{i=0}^{\infty} \subset (0,1]$, $\{\Delta_i\}_{i=0}^{\infty} \subset (0,1]$ satisfy*

$$\lim_{i \to \infty} \delta_i = 0, \ \lim_{i \to \infty} \Delta_i = 0,$$

a sequence $\{x_i\}_{i=0}^{\infty} \subset C$ is bounded, $\{l_i\}_{i=0}^{\infty} \subset X$, and for each integer $i \geq 0$,

$$l_i \in E(x_i, \Delta_i, \Delta_i),$$

$$x_{i+1} \in \{x_i + \alpha l_i : \ \alpha \in [0,1]\},$$

and

$$f(x_{i+1}) \leq f(x_i + \beta l_i) + \delta_i, \ \beta \in [0,1].$$

Then

$$\lim_{i \to \infty} f(x_i) = \inf(f, C).$$

The next result is also related to algorithm $\mathcal{A}1$. We assume that the sequence of iterates $\{x_t\}_{t=0}^{\infty}$ has a bounded subsequence and define $x_{t+1} = x_t + \alpha_t l_t$ only if $f(x_t + \alpha_t l_t) \leq f(x_t)$; otherwise, $x_{t+1} = x_t$. It is proved in Sect. 2.7.

Theorem 2.5. *Assume that $\{\delta_i\}_{i=0}^{\infty} \subset (0,1]$, $\{\Delta_i\}_{i=0}^{\infty} \subset (0,1]$ satisfy*

$$\lim_{i \to \infty} \delta_i = 0, \ \lim_{i \to \infty} \Delta_i = 0,$$

a sequence $\{x_i\}_{i=0}^{\infty} \subset C$ satisfies

$$\liminf_{i \to \infty} \|x_i\| < \infty,$$

and for each integer $i \geq 0$,

$$l_i \in E(x_i, \Delta_i, \Delta_i), \ \alpha_i \in [0, 1],$$

$$f(x_i + \alpha_i l_i) \leq f(x_i + \beta l_i) + \delta_i, \ \beta \in [0, 1],$$

$$x_{i+1} = x_i + \alpha_i l_i \ \text{if} \ f(x_i + \alpha_i l_i) \leq f(x_i); \ \text{otherwise} \ x_{i+1} = x_i.$$

Then

$$\lim_{i \to \infty} f(x_i) = \inf(f, C).$$

The following theorem is related to algorithm $\mathcal{A}2$. We assume that the sequence of iterates $\{x_t\}_{t=0}^{\infty}$ is bounded and define $x_{t+1} = x_t + \alpha_t l_t$ only if $f(x_t + \alpha_t l_t) \leq f(x_t)$; otherwise, $x_{t+1} = x_t$. It is proved in Sect. 2.8.

Theorem 2.6. *Assume that $\{\alpha_t\}_{t=0}^{\infty} \subset (0, 1]$ satisfies*

$$\lim_{t \to \infty} \alpha_t = 0, \ \sum_{t=0}^{\infty} \alpha_t = \infty,$$

$\{\Delta_i\}_{i=0}^{\infty} \subset (0, 1]$ *satisfies*

$$\lim_{i \to \infty} \Delta_i = 0,$$

a sequence $\{x_i\}_{i=0}^{\infty} \subset C$ is bounded, and for each integer $i \geq 0$,

$$l_i \in E(x_i, \Delta_i, \Delta_i),$$

if

$$f(x_i + \alpha_i l_i) \leq f(x_i),$$

then

$$x_{i+1} = x_i + \alpha_i l_i;$$

otherwise, $x_{i+1} = x_i$. Then

$$\lim_{i \to \infty} f(x_i) = \inf(f, C).$$

The next result is also related to algorithm $\mathcal{A}2$. We assume that the sequence of iterates $\{x_t\}_{t=0}^{\infty}$ is bounded. The theorem is proved in Sect. 2.9.

Theorem 2.7. *Assume that $\{\alpha_i\}_{i=0}^{\infty} \subset (0, 1]$ satisfies*

$$\lim_{i \to \infty} \alpha_i = 0, \ \sum_{i=0}^{\infty} \alpha_i = \infty,$$

$\{\Delta_i\}_{i=0}^{\infty} \subset (0, 1]$ *satisfies*

$$\lim_{i \to \infty} \Delta_i = 0,$$

a sequence $\{x_i\}_{i=0}^\infty \subset C$ *is bounded, and for each integer* $i \geq 0$,

$$l_i \in E(x_i, \Delta_i, \Delta_i),$$

$$x_{i+1} = x_i + \alpha_i l_i.$$

Then

$$\lim_{i\to\infty} f(x_i) = \inf(f, C).$$

The following theorem is related to algorithm $\mathcal{A}2$. We assume that the objective function f satisfies the growth condition and that the sequence of iterates $\{x_t\}_{t=0}^\infty$ has bounded subsequence. It is proved in Sect. 2.10.

Theorem 2.8. *Assume that*

$$\lim_{\|x\|\to\infty,\, x\in U} f(x) = \infty,$$

$\{\alpha_i\}_{i=0}^\infty \subset (0, 1]$ *satisfies*

$$\lim_{i\to\infty} \alpha_i = 0, \quad \sum_{i=0}^\infty \alpha_i = \infty,$$

and $K, \epsilon > 0$. *Then there exists* $\Delta \in (0, 1]$ *such that for each sequence* $\{x_i\}_{i=0}^\infty \subset C$ *and each sequence* $\{l_i\}_{i=0}^\infty \subset X$ *that satisfy*

$$\liminf_{i\to\infty} \|x_i\| < K,$$

and for each integer $i \geq 0$,

$$\xi_i \in E(x_i, \Delta, \Delta)$$

and

$$x_{i+1} = x_i + \alpha_i l_i,$$

the inequality

$$f(x_i) \leq \inf(f, C) + \epsilon$$

holds for all sufficiently large natural numbers i.

The following theorem is related to algorithm $\mathcal{A}2$. We assume that the objective function f satisfies the growth condition and that the sequence of step sizes is bounded by a sufficiently small constant. It is proved in Sect. 2.11.

Theorem 2.9. *Assume that*

$$\lim_{\|x\|\to\infty,\, x\in U} f(x) = \infty$$

and $K, \epsilon > 0$. Then there exist $\alpha, \Delta \in (0, 1)$ such that the following assertion holds.

Assume that $\{\alpha_i\}_{i=0}^{\infty} \subset (0, \alpha]$ satisfies

$$\sum_{i=0}^{\infty} \alpha_i = \infty.$$

Then there exists a natural number N_0 such that for each sequence $\{x_i\}_{i=0}^{\infty} \subset C$ and each sequence $\{l_i\}_{i=0}^{\infty} \subset X$ that satisfy

$$\|x_0\| \leq K$$

and for each integer $i \geq 0$,

$$l_i \in E(x_i, \Delta, \Delta)$$

and

$$x_{i+1} = x_i + \alpha_i l_i,$$

the inequality

$$f(x_i) \leq \inf(f, C) + \epsilon$$

holds for all integers $i \geq N_0$.

The final theorem of this section is related to algorithm $\mathcal{A}3$. We assume that the objective function f satisfies the growth condition. It is proved in Sect. 2.12.

Theorem 2.10. *Assume that*

$$\lim_{\|x\| \to \infty, \, x \in U} f(x) = \infty$$

and $K, \epsilon > 0$. *Then there exist* $\beta_0 \in (0, 1]$ *and* $\Delta > 0$ *such that for each* $\beta_1 \in (0, \beta_0]$ *there exists a natural number* N_0 *such that the following assertion holds.*

Assume that $n \geq N_0$ *is an integer,* $\{x_i\}_{i=0}^{n} \subset C$, $\{l_i\}_{i=0}^{n-1} \subset X$,

$$\alpha_i \in [\beta_1, \beta_0], \, i = 0, \ldots, n-1,$$

$$\|x_0\| \leq K,$$

and for each integer $i \in \{0, \ldots, n-1\}$,

$$l_i \in E(x_i, \Delta, \Delta)$$

and

$$x_{i+1} = x_i + \alpha_i l_i.$$

Then the inequality

$$f(x_i) \leq \inf(f, C) + \epsilon$$

holds for all integers $i \in \{N_0, \ldots, n\}$.

2.4 Proof of Theorem 2.2

We may assume without loss of generality that $\epsilon < 1$, $K > 2$ and that

$$\{x \in U : f(x) \leq \inf(f, C) + 4\} \subset B(0, K - 2). \tag{2.37}$$

Fix

$$K_0 > \sup\{f(x) : x \in B(0, K + 1) \cap U\}. \tag{2.38}$$

Set

$$E_0 = \{x \in U : f(x) \leq K_0 + 1\}. \tag{2.39}$$

Clearly, E_0 is a bounded set. Set

$$K_1 = \sup\{\|x\| : x \in E_0\}. \tag{2.40}$$

There exists $L > 1$ such that

$$|f(z_1) - f(z_2)| \leq L\|z_1 - z_2\|, \ z_1, z_2 \in B(0, K + 1) \cap U. \tag{2.41}$$

Lemma 2.1 and (2.37)–(2.40) imply that there exist $\bar{\alpha}, \gamma, \Delta, \bar{\delta} \in (0, 1)$ such that the following property holds:

(a) For each $x \in B(0, K_1) \cap C$ satisfying

$$f(x) \geq \inf(f, C) + \epsilon/2,$$

each $\xi \in E(x, \bar{\delta}, \Delta)$, and each $\beta \in (0, \bar{\alpha}]$, the inequality

$$f(x) - f(x + \beta\xi) \geq \beta\gamma$$

holds.

Set

$$\alpha = \min\{\bar{\alpha}, \ (c_* + 1)^{-1}L^{-1}\epsilon/4\}, \tag{2.42}$$

$$\delta = \min\{\bar{\delta}, \ \bar{\alpha}\gamma/2, \ \epsilon/4\} \tag{2.43}$$

and choose an integer

$$N_0 > 2(K_0 - \inf(f, C))(\bar{\alpha}\gamma)^{-1}. \tag{2.44}$$

Assume that $\{x_i\}_{i=0}^\infty \subset C$, $\{\xi_i\}_{i=0}^\infty \subset X$,

$$\|x_0\| \leq K \tag{2.45}$$

and that for each integer $i \geq 0$,

$$\xi_i \in E(x_i, \delta, \Delta), \tag{2.46}$$

$$x_{i+1} \in \{x_i + \alpha\xi_i : \alpha \in [0,1]\}, \tag{2.47}$$

and

$$f(x_{i+1}) \le f(x_i + \alpha\xi_i) + \delta, \ \alpha \in [0,1]. \tag{2.48}$$

In order to complete the proof of the theorem, it is sufficient to show that

$$f(x_i) \le \inf(f,C) + \epsilon$$

for all integers $i \ge N_0$. By (2.47), for each integer $i \ge 0$, there exists $\alpha_i \in [0,1]$ such that

$$x_{i+1} = x_i + \alpha_i\xi_i. \tag{2.49}$$

We show that there exists an integer $j \in \{0, \dots, N_0\}$ such that

$$f(x_j) \le \inf(f,C) + \epsilon.$$

Assume the contrary. Then

$$f(x_i) > \inf(f,C) + \epsilon, \ i = 0, \dots, N_0. \tag{2.50}$$

In particular,

$$f(x_0) > \inf(f,C) + \epsilon.$$

Together with property (a), the inequality $K_1 > K$ and Eqs. (2.42), (2.43), (2.45), (2.46), (2.48)–(2.50), this implies that

$$f(x_1) = f(x_0 + \alpha_0\xi_0) \le f(x_0 + \bar\alpha\xi_0) + \delta$$

$$\le f(x_0) - \bar\alpha\gamma + \delta \le f(x_0) - \bar\alpha\gamma/2. \tag{2.51}$$

Equations (2.38), (2.40), (2.45), and (2.51) imply that

$$f(x_1) < K_0, \ \|x_1\| \le K_1, \ f(x_1) \le f(x_0) - \bar\alpha\gamma/2. \tag{2.52}$$

By induction, we show that for each integer $i \in \{0, \dots, N_0\}$,

$$\|x_i\| \le K_1 \text{ and } f(x_i) \le f(x_{i-1}) - \bar\alpha\gamma2. \tag{2.53}$$

In view of (2.52), Eq. (2.53) holds for $i = 1$. Assume that $q \in \{1, \dots, N_0 - 1\}$ and (2.53) holds for $i = 1, \dots, q$. It follows from (2.38)–(2.40), (2.52), and (2.53) that

$$f(x_q) \le f(x_0) \le K_0, \ \|x_q\| \le K_1. \tag{2.54}$$

Property (a) and equations (2.40), (2.42), (2.43), (2.46), (2.48)–(2.50), and (2.54) imply that

$$f(x_{q+1}) = f(x_q + \alpha_q\xi_q) \le f(x_q + \bar\alpha_q\xi_q) + \delta$$

$$\leq f(x_q) - \bar{\alpha}\gamma + \delta \leq f(x_q) - \bar{\alpha}\gamma/2. \tag{2.55}$$

In view of (2.39), (2.40), (2.54), and (2.55),

$$f(x_{q+1}) \leq K_0, \ \|x_{q+1}\| \leq K_1.$$

Together with (2.55), this implies that (2.53) holds for all $i = 1, \ldots, q + 1$. Therefore by induction, we showed that (2.53) holds for all integers $i = 0, \ldots, N_0$. By (2.38), (2.45), and (2.53),

$$K_0 - \inf(f, C) \geq f(x_0) - f(x_{N_0})$$

$$= \sum_{i=0}^{N_0-1} (f(x_i) - f(x_{i+1})) \geq N_0 \bar{\alpha}\gamma/2$$

and

$$N_0 \leq 2(K_0 - \inf(f, C))(\bar{\alpha}\gamma)^{-1}.$$

This contradicts (2.44). The contradiction we have reached proves that there exists

$$j_0 \in \{0, \ldots, N_0\} \tag{2.56}$$

such that

$$f(x_{j_0}) \leq \inf(f, C) + \epsilon. \tag{2.57}$$

By (2.37) and (2.57),

$$\|x_{j_0}\| \leq K. \tag{2.58}$$

By induction, we show that for each integer $i \geq j_0$,

$$\|x_i\| \leq K, \ f(x_i) \leq \inf(f, C) + \epsilon. \tag{2.59}$$

In view of (2.57) and (2.58), Eq. (2.59) is true for $i = j_0$.

Assume that $i \geq j_0$ is an integer and (2.59) is true. There are two cases:

$$f(x_i) \geq \inf(f, C) + \epsilon/2; \tag{2.60}$$

$$f(x_i) < \inf(f, C) + \epsilon/2. \tag{2.61}$$

Assume that (2.60) holds. Property (a), (2.38)–(2.40), (2.42), (2.43), (2.46), (2.48), (2.49), (2.59), and (2.60) imply that

$$f(x_{i+1}) = f(x_i + \alpha_i \xi_i) \leq f(x_i + \bar{\alpha}\xi_i) + \delta$$

$$\leq f(x_i) - \bar{\alpha}\gamma + \delta \leq f(x_i) - \bar{\alpha}\gamma/2 \leq \inf(f, C) + \epsilon. \tag{2.62}$$

Assume that (2.61) holds. By (2.42), (2.48), and (2.49),

$$f(x_{i+1}) = f(x_i + \alpha_i \xi_i) \leq f(x_i + \alpha \xi_i) + \delta. \tag{2.63}$$

In view of (2.43) and (2.46),

$$\|\xi_i\| \leq c_* + \delta \leq c_* + 1. \tag{2.64}$$

By (2.42) and (2.64),

$$\|\alpha \xi_i\| \leq (c_* + 1)\alpha \leq 1. \tag{2.65}$$

Equations (2.59) and (2.65) imply that

$$\|x_i + \alpha \xi_i\| \leq K + 1. \tag{2.66}$$

It follows from (2.41), (2.42), (2.59), (2.64), and (2.66) that

$$|f(x_i) - f(x_i + \alpha \xi_i)| \leq L\|\alpha \xi_i\| \leq L(c_* + 1)\alpha \leq \epsilon/4. \tag{2.67}$$

It follows from (2.43), (2.63), and (2.67) that

$$f(x_{i+1}) \leq f(x_i + \alpha \xi_i) + \delta \leq f(x_i) + \epsilon/4 + \delta \leq \inf(f, C) + \epsilon.$$

Thus in both cases (see (2.62)),

$$f(x_{i+1}) \leq \inf(f, C) + \epsilon.$$

Together with (2.37), this implies that

$$\|x_{i+1}\| \leq K.$$

Therefore, we showed by induction that (2.59) holds for all integers $i \geq j_0$. Theorem 2.2 is proved.

2.5 Proof of Theorem 2.3

We may assume without loss of generality that $\epsilon < 1$, $K > 2$ and that

$$\{x \in U : f(x) \leq \inf(f, C) + 4\} \subset B(0, K - 2). \tag{2.68}$$

Fix

$$K_0 > \sup\{f(x) : x \in B(0, K + 1) \cap U\}. \tag{2.69}$$

Set

$$E_0 = \{x \in U : f(x) \leq K_0 + 1\}. \tag{2.70}$$

Clearly, E_0 is a bounded set. Fix

$$K_1 > \sup\{\|x\| : x \in E_0\}. \tag{2.71}$$

Note that

$$\lim_{i \to \infty} \alpha_i = 0, \quad \sum_{i=0}^{\infty} \alpha_i = \infty. \tag{2.72}$$

Lemma 2.1 and (2.68)–(2.71) imply that there exist $\bar{\alpha}, \gamma, \Delta, \delta \in (0, 1)$ such that the following property holds:

(a) For each $x \in B(0, K_1) \cap C$ satisfying

$$f(x) \geq \inf(f, C) + \epsilon/4,$$

each $\xi \in E(x, \delta, \Delta)$, and each $\beta \in (0, \bar{\alpha}]$, the inequality

$$f(x) - f(x + \beta\xi) \geq \beta\gamma$$

holds.

In view of (2.72), there exists a natural number N_1 such that

$$\alpha_i < \bar{\alpha} \text{ for all integers } i \geq N_1 \tag{2.73}$$

and a natural number $N_0 > N_1 + 2$ such that

$$\sum_{i=N_1}^{N_0-1} \alpha_i > \gamma^{-1}(K_0 - \inf(f, C)). \tag{2.74}$$

Assume that $\{x_i\}_{i=0}^{\infty} \subset C$, $\{\xi_i\}_{i=0}^{\infty} \subset X$,

$$\|x_0\| \leq K \tag{2.75}$$

and that for each integer $i \geq 0$,

$$\xi_i \in E(x_i, \delta, \Delta), \tag{2.76}$$

$$x_{i+1} = x_i + \alpha\xi_i \text{ if } f(x_i + \alpha_i\xi_i) < f(x_i); \text{ otherwise, } x_{i+1} = x_i. \tag{2.77}$$

In view of (2.77), for each integer $i \geq 0$,

$$f(x_{i+1}) \leq f(x_i) \leq f(x_0). \tag{2.78}$$

Equations (2.69)–(2.71), (2.75), and (2.78) imply that for each integer $i \geq 0$,

$$f(x_i) \leq K_0, \|x_i\| \leq K_1. \tag{2.79}$$

In order to complete the proof of the theorem, it is sufficient to show that

$$f(x_{N_0}) \leq \inf(f, C) + \epsilon.$$

Assume the contrary. Then for every integer $i \in \{0, \ldots, N_0\}$,

$$f(x_i) > \inf(f, C) + \epsilon. \tag{2.80}$$

By (2.73), (2.76), (2.79), and (2.80), for every integer $i \in [N_1, N_0]$,

$$f(x_i) - f(x_i + \alpha_i\xi_i) \geq \alpha_i\gamma. \tag{2.81}$$

In view of (2.77) and (2.81),

$$x_{i+1} = x_i + \alpha_i \xi_i,$$

$$f(x_i) - f(x_{i+1}) \geq \alpha_i \gamma. \tag{2.82}$$

Equations (2.79) and (2.82) imply that

$$K_0 - \inf(f, C) \geq f(x_{N_1}) - f(x_{N_0})$$

$$= \sum_{i=N_1}^{N_0-1} (f(x_i) - f(x_{i+1})) \geq \gamma \sum_{i=N_1}^{N_0-1} \alpha_i$$

and

$$\sum_{i=N_1}^{N_0-1} \alpha_i \leq \gamma^{-1}(K_0 - \inf(f, C)).$$

This contradicts (2.74). The contradiction we have reached proves Theorem 2.3.

2.6 Proof of Theorem 2.4

Note that

$$\lim_{i \to \infty} \delta_i = 0, \ \lim_{i \to \infty} \Delta_i = 0. \tag{2.83}$$

There exists $K > 1$ such that

$$\|x_i\| \leq K, \ i = 0, 1, \ldots \tag{2.84}$$

In view of (2.84), there exists $K_0 > 0$ such that

$$|f(x_i)| < K_0, \ i = 0, 1, \ldots \tag{2.85}$$

There exists $L_0 > 1$ such that

$$|f(z_1) - f(z_2)| \leq L_0\|z_1 - z_2\|, \ z_1, z_2 \in B(0, K+1) \cap U. \tag{2.86}$$

Let $\epsilon \in (0, 1)$. In order to complete the proof of the theorem, it is sufficient to show that for all sufficiently large natural numbers i, we have

$$f(x_i) \leq \inf(f, C) + \epsilon.$$

Note that for all integers $i \geq 0$,

$$l_i \in E(x_i, \Delta_i, \Delta_i), \tag{2.87}$$

$$f(x_{i+1}) \leq f(x_i + \beta l_i) + \delta_i, \ \beta \in [0, 1]. \tag{2.88}$$

Lemma 2.1 implies that there exist $\bar{\alpha}, \gamma, \Delta, \delta \in (0, 1)$ such that the following property holds:

(a) For each $x \in B(0, K) \cap C$ satisfying

$$f(x) \geq \inf(f, C) + \epsilon/4,$$

each $\xi \in E(x, \delta, \Delta)$, and each $\beta \in (0, \bar{\alpha}]$, the inequality

$$f(x) - f(x + \beta\xi) \geq \beta\gamma$$

holds.

We may assume without loss of generality that

$$\gamma \leq \epsilon/4. \tag{2.89}$$

By (2.83), there exists a natural number $N_1 > 2$ such that for every integer $i \geq N_1$,

$$\Delta_i, \delta_i < \min\{\delta, \Delta, \bar{\alpha}\gamma/2\}. \tag{2.90}$$

Set

$$\alpha = \min\{\bar{\alpha}, \ (8c_*)^{-1}(L_0 + 1)^{-1}\epsilon\}. \tag{2.91}$$

Choose a natural number $N_0 > N_1 + 2$ such that

$$N_0 > N_1 + 4K_0(\bar{\alpha}\gamma)^{-1}. \tag{2.92}$$

We show that there exists an integer $j \in \{N_1, \ldots, N_0\}$ such that

$$f(x_j) \leq \inf(f, C) + \epsilon/4.$$

Assume the contrary. Then

$$f(x_i) > \inf(f, C) + \epsilon/4, \ i = N_1, \ldots, N_0. \tag{2.93}$$

Let

$$i \in \{N_1, \ldots, N_0\}. \tag{2.94}$$

Property (a) and Eqs. (2.84), (2.87), (2.90), (2.93), and (2.94) imply that

$$f(x_i) - f(x_i + \bar{\alpha}l_i) \geq \bar{\alpha}\gamma. \tag{2.95}$$

In view of (2.88), (2.90), (2.91), and (2.95),

$$f(x_{i+1}) \leq f(x_i + \bar{\alpha}l_i) + \delta_i \leq f(x_i) - \bar{\alpha}\gamma + \bar{\alpha}\gamma/2$$

$$\leq f(x_i) - \bar{\alpha}\gamma/2. \tag{2.96}$$

By (2.85) and (2.96),

$$2K_0 \geq f(x_{N_1}) - f(x_{N_0})$$

$$= \sum_{i=N_1}^{N_0-1} (f(x_i) - f(x_{i+1})) \geq (N_0 - N_1)\bar{\alpha}\gamma/2$$

and

$$N_0 \leq N_1 + 4K_0(\bar{\alpha}\gamma)^{-1}.$$

This contradicts (2.92). The contradiction we have reached proves that there exists

$$j \in \{N_1, \ldots, N_0\}$$

such that

$$f(x_j) < \inf(f, C) + \epsilon/4. \tag{2.97}$$

Assume that $i \geq j$ is an integer and that

$$f(x_i) \leq \inf(f, C) + \epsilon. \tag{2.98}$$

We show that

$$f(x_{i+1}) \leq \inf(f, C) + \epsilon$$

too.

There are two cases:

$$f(x_i) > \inf(f, C) + \epsilon/4; \tag{2.99}$$

$$f(x_i) \leq \inf(f, C) + \epsilon/4. \tag{2.100}$$

Assume that (2.99) holds. Property (a), the inequality $j \geq N_1$, (2.84), (2.87), (2.88), (2.90), and (2.98) imply that

$$f(x_{i+1}) \leq f(x_i + \bar{\alpha}l_i) + \delta_i$$

$$\leq f(x_i) - \bar{\alpha}\gamma + \bar{\alpha}\gamma/2 \leq f(x_i) - \bar{\alpha}\gamma/2 \leq f(x_i) \leq \inf(f, C) + \epsilon.$$

Assume that (2.100) holds. By (2.88), (2.90), (2.91) and the inequality $i \geq j \geq N_1$,

$$f(x_{i+1}) \leq f(x_i + \alpha l_i) + \delta_i \leq f(x_i + \alpha l_i) + \bar{\alpha}\gamma/2. \tag{2.101}$$

By (2.87) and (2.91),

$$\|\alpha l_i\| \leq (c_* + 1)\alpha \leq 2c_*\alpha \leq 1. \tag{2.102}$$

Equations (2.84) and (2.102) imply that

$$\|x_i + \alpha l_i\| \leq K + 1. \tag{2.103}$$

It follows from (2.84), (2.86), (2.90), (2.102), and (2.103) that

$$|f(x_i) - f(x_i + \alpha l_i)| \le L_0 \|\alpha l_i\| \le 2\alpha L_0 c_* \le \epsilon/4. \tag{2.104}$$

By (2.89), (2.100), (2.101), and (2.104),

$$f(x_{i+1}) \le f(x_i) + \epsilon/4 + \bar{\alpha}\gamma/2 \le \inf(f, C) + 3\epsilon/4.$$

Thus in both cases

$$f(x_{i+1}) \le \inf(f, C) + \epsilon.$$

Therefore, we showed by induction that

$$f(x_i) \le \inf(f, C) + \epsilon$$

for all integers $i \ge j$. Theorem 2.4 is proved.

2.7 Proof of Theorem 2.5

Recall that

$$\lim_{i \to \infty} \Delta_i = 0, \ \lim_{i \to \infty} \delta_i = 0, \tag{2.105}$$

for each integer $i \ge 0$,

$$l_i \in E(x_i, \Delta_i, \Delta_i), \tag{2.106}$$

$$f(x_i + \alpha_i l_i) \le f(x_i + \beta l_i) + \delta_i, \ \beta \in [0, 1], \tag{2.107}$$

if $f(x_i + \alpha_i l_i) \le f(x_i)$, then $x_{i+1} = x_i + \alpha_i l_i$; otherwise, $x_{i+1} = x_i$. (2.108)
There exist a subsequence $\{x_{n_p}\}_{p=1}^{\infty}$ and $K > 1$ such that

$$\|x_{n_p}\| \le K, \ p = 1, 2, \ldots \tag{2.109}$$

In view of (2.108), for each integer $i \ge 0$,

$$f(x_{i+1}) \le f(x_i) \le f(x_0). \tag{2.110}$$

Let $\epsilon \in (0, 1)$. In view of (2.110), in order to complete the proof of the theorem, it is sufficient to show that there exists an integer $n \ge 0$ such that

$$f(x_n) \le \inf(f, C) + \epsilon.$$

Lemma 2.1 implies that there exist $\bar{\alpha}, \gamma, \Delta, \delta \in (0, 1)$ such that the following property holds:

(a) For each $x \in B(0, K) \cap C$ satisfying

$$f(x) \geq \inf(f, C) + \epsilon/4,$$

each $l \in E(x, \delta, \Delta)$, and each $\beta \in (0, \bar{\alpha}]$, the inequality

$$f(x) - f(x + \beta l) \geq \beta \gamma$$

holds.

We may assume without loss of generality that

$$\gamma \leq \epsilon/4. \tag{2.111}$$

By (2.105), there exists a natural number p_1 such that for each integer $i \geq n_{p_1}$,

$$\Delta_i, \delta_i < \min\{\delta, \Delta, \bar{\alpha}\gamma/4\}. \tag{2.112}$$

Choose a natural number

$$p_2 > p_1 + 2(\bar{\alpha}\gamma)^{-1}(f(x_0) - \inf(f, C)). \tag{2.113}$$

We show that there exists an integer $j \in \{n_{p_1}, \ldots, n_{p_2}\}$ such that

$$f(x_j) \leq \inf(f, C) + \epsilon/4.$$

Assume the contrary. Then

$$f(x_i) > \inf(f, C) + \epsilon/4, \ i = n_{p_1}, \ldots, n_{p_2}. \tag{2.114}$$

Let

$$p \in \{p_1, \ldots, p_2\}. \tag{2.115}$$

Property (a) and Eqs. (2.106), (2.109), (2.212), and (2.214) imply that

$$f(x_{n_p}) - f(x_{n_p} + \bar{\alpha}l_{n_p}) \geq \bar{\alpha}\gamma. \tag{2.116}$$

In view of (2.107), (2.108), (2.212), (2.115), and (2.116),

$$f(x_{n_p} + \alpha_{n_p}l_{n_p}) \leq f(x_{n_p} + \bar{\alpha}l_{n_p}) + \delta_{n_p}$$

$$\leq f(x_{n_p}) - \bar{\alpha}\gamma + \delta_{n_p} \leq f(x_{n_p}) - \bar{\alpha}\gamma/2,$$

$$x_{n_p+1} = x_{n_p} + \alpha_{n_p}l_{n_p},$$

$$f(x_{n_p+1}) \leq f(x_{n_p}) - \bar{\alpha}\gamma/2. \tag{2.117}$$

It follows from (2.110) and (2.117) that

$$f(x_0) - \inf(f, C) \geq f(x_{n_{p_1}}) - f(x_{n_{p_2}})$$

$$= \sum_{p=p_1}^{p_2-1} (f(x_{n_p}) - f(x_{n_{p+1}})) \geq (p_2 - p_1)\bar{\alpha}\gamma/2$$

and

$$p_2 \leq p_1 + 2(\bar{\alpha}\gamma)^{-1}(f(x_0) - \inf(f, C)).$$

This contradicts (2.113). The contradiction we have reached proves that there exists

$$j \in \{n_{p_1}, \ldots, n_{p_2}\}$$

such that

$$f(x_j) \leq \inf(f, C) + \epsilon.$$

Theorem 2.5 is proved.

2.8 Proof of Theorem 2.6

There exists $K > 1$ such that

$$\|x_i\| \leq K, \ i = 0, 1, \ldots \tag{2.118}$$

Let $\epsilon \in (0, 1)$. In order to complete the proof of the theorem, it is sufficient to show that there exists an integer $i \geq 0$ such that

$$f(x_i) \leq \inf(f, C) + \epsilon.$$

Lemma 2.1 implies that there exist $\bar{\alpha}, \gamma, \Delta, \delta \in (0, 1)$ such that the following property holds:

(a) For each $x \in B(0, K) \cap C$ satisfying

$$f(x) \geq \inf(f, C) + \epsilon/4,$$

each $l \in E(x, \delta, \Delta)$, and each $\beta \in (0, \bar{\alpha}]$, the inequality

$$f(x) - f(x + \beta\xi) \geq \beta\gamma$$

holds.

In view of (2.118), there exists $K_1 > 0$ such that

$$|f(x_i)| \leq K_1, \ i = 0, 1, \ldots \tag{2.119}$$

There exist a natural number $N_1 > 1$ such that for each integer $i \geq N_1$,

$$\Delta_i < \min\{\delta, \Delta\} \tag{2.120}$$

and a natural number $N_0 > N_1 + 2$ such that

$$\sum_{i=N_1}^{N_0-1} > 2K_1\gamma^{-1}. \tag{2.121}$$

By definition,

$$f(x_{i+1}) \le f(x_i), \ i = 0, 1, \ldots \tag{2.122}$$

In order to prove the theorem, it is sufficient to show that

$$f(x_{N_0}) \le \inf(f, C) + \epsilon.$$

Assume the contrary. Then for all integers $i = 0, \ldots, N_0$,

$$f(x_i) > \inf(f, C) + \epsilon. \tag{2.123}$$

Note that for every integer $i \ge 0$,

$$l_i \in E(x_i, \Delta_i, \Delta_i). \tag{2.124}$$

Property (a) and Eqs. (2.118), (2.120), (2.123), and (2.124) imply that for each integer $i \in [N_1, N_0)$,

$$f(x_i + \alpha_i l_i) \le f(x_i) - \alpha_i \gamma. \tag{2.125}$$

In view of (2.125), for each integer $i \in [N_1, N_0)$,

$$x_{i+1} = x_i + \alpha_i x_i, \ f(x_{i+1}) \le f(x_i) - \alpha_i \gamma. \tag{2.126}$$

By (2.119) and (2.126),

$$2K_1 \ge f(x_{N_1}) - f(x_{N_0})$$

$$= \sum_{i=N_1}^{N_0-1} (f(x_i) - f(x_{i+1})) \ge \gamma \sum_{i=N_1}^{N_0-1} \alpha_i$$

and

$$\sum_{i=N_1}^{N_0-1} \alpha_i \le 2K_1\gamma^{-1}.$$

This contradicts (2.121). The contradiction we have reached completes the proof of Theorem 2.6.

2.9 Proof of Theorem 2.7

There exists $K > 1$ such that

$$\|x_i\| \le K, \; i = 0, 1, \ldots \tag{2.127}$$

Let $\epsilon \in (0, 1)$. In order to complete the proof of the theorem, it is sufficient to show that for all sufficiently large natural numbers i, we have

$$f(x_i) \le \inf(f, C) + \epsilon.$$

Lemma 2.1 implies that there exist $\bar{\alpha}, \gamma, \Delta, \delta \in (0, 1)$ such that the following property holds:

(a) For each $x \in B(0, K) \cap C$ satisfying

$$f(x) \ge \inf(f, C) + \epsilon/4,$$

each $l \in E(x, \delta, \Delta)$, and each $\beta \in (0, \bar{\alpha}]$, the inequality

$$f(x) - f(x + \beta l) \ge \beta \gamma$$

holds.

There exist $L_0 \ge 1$ and $K_0 > 0$ such that

$$|f(z_1) - f(z_2)| \le L_0 \|z_1 - z_2\|, \; z_1, z_2 \in B(0, K+1) \cap U \tag{2.128}$$

and

$$|f(x)| < K_0, \; x \in B(0, K+1) \cap U. \tag{2.129}$$

Since

$$\lim_{i \to \infty} \alpha_i = 0, \; \lim_{i \to \infty} \Delta_i = 0,$$

there exists a natural number $N_1 > 1$ such that for each integer $i \ge N_1$,

$$\alpha_i < \min\{\bar{\alpha}, (8c_* L_0)^{-1}\epsilon\}, \; \Delta_i < \min\{\delta, \Delta\}. \tag{2.130}$$

Since

$$\sum_{i=0}^{\infty} \alpha_i = \infty,$$

there exists an integer $N_0 > N_1 + 2$ such that

$$\sum_{i=N_1+1}^{N_0-1} \alpha_i > 2K_0 \gamma^{-1}. \tag{2.131}$$

We show that there exists an integer $j \in \{N_1 + 1, \ldots, N_0\}$ such that

$$f(x_j) \le \inf(f, C) + \epsilon/4.$$

Assume the contrary. Then

$$f(x_i) > \inf(f, C) + \epsilon/4, \ i = N_1 + 1, \ldots, N_0. \tag{2.132}$$

Note that for all integers $i \geq 0$,

$$l_i \in E(x_i, \Delta_i, \Delta_i). \tag{2.133}$$

Let

$$i \in \{N_1 + 1, \ldots, N_0\}. \tag{2.134}$$

Property (a) and Eqs. (2.127), (2.130), and (2.132)–(2.134) imply that

$$f(x_i) - f(x_{i+1}) = f(x_i) - f(x_i + \alpha_i l_i) \geq \alpha_i \gamma. \tag{2.135}$$

By (2.129) and (2.135),

$$2K_0 \geq f(x_{N_1+1}) - f(x_{N_0})$$

$$= \sum_{i=N_1+1}^{N_0-1} (f(x_i) - f(x_{i+1})) \geq \gamma \sum_{i=N_1+1}^{N_0-1} \alpha_i$$

and

$$\sum_{i=N_1+1}^{N_0-1} \alpha_i \leq 2K_0 \gamma^{-1}.$$

This contradicts (2.131). The contradiction we have reached proves that there exists

$$j \in \{N_1 + 1, \ldots, N_0\} \tag{2.136}$$

such that

$$f(x_j) \leq \inf(f, C) + \epsilon/4.$$

Assume that $i \geq j$ is an integer and that

$$f(x_i) \leq \inf(f, C) + \epsilon. \tag{2.137}$$

We show that

$$f(x_{i+1}) \leq \inf(f, C) + \epsilon$$

too. There are two cases:

$$f(x_i) > \inf(f, C) + \epsilon/4; \tag{2.138}$$

$$f(x_i) \leq \inf(f, C) + \epsilon/4. \tag{2.139}$$

Assume that (2.138) holds. Property (a), (2.127), (2.130), (2.133), (2.138) and the inequality $i \geq j > N_1$ imply that

$$f(x_i) - f(x_{i+1}) = f(x_i) - f(x_i + \alpha_i l_i) \geq \alpha_i \gamma.$$

Together with (2.137), this implies that

$$f(x_{i+1}) \leq f(x_i) \leq \inf(f, C) + \epsilon.$$

Assume that (2.139) holds. By (2.130) and (2.133),

$$\|x_{i+1} - x_i\| \leq \alpha_i \|l_i\| \leq (c_* + 1)\alpha_i \leq 2c_* \alpha_i. \tag{2.140}$$

Equations (2.127), (2.128), (2.130), and (2.140) imply that

$$|f(x_i) - f(x_{i+1})| \leq L_0 \|x_i - x_{i+1}\| \leq 2\alpha_i L_0 c_* \leq \epsilon/4.$$

Together with (2.139), this implies that

$$f(x_{i+1}) \leq f(x_i) + \epsilon/4 \leq \inf(f, C) + \epsilon.$$

Thus in both cases

$$f(x_{i+1}) \leq \inf(f, C) + \epsilon.$$

This completes the proof of Theorem 2.7.

2.10 Proof of Theorem 2.8

We may assume without loss of generality that

$$\epsilon < 1, \ K > 2. \tag{2.141}$$

Note that

$$\lim_{\|x\| \to \infty, \, x \in U} f(x) = \infty. \tag{2.142}$$

In view of (2.142), we may assume without loss of generality that

$$\{x \in U : f(x) \leq \inf(f, C) + 4\} \subset B(0, K - 2). \tag{2.143}$$

Fix

$$K_0 > \sup\{f(x) : x \in B(0, K + 1) \cap U\}. \tag{2.144}$$

Set

$$E_0 = \{x \in U : f(x) \leq K_0 + 1\}. \tag{2.145}$$

Clearly, E_0 is a bounded set. Fix

$$K_1 > \sup\{\|x\| : x \in E_0\}. \tag{2.146}$$

There exists $L_0 \geq 1$ such that

$$|f(z_1) - f(z_2)| \leq L_0 \|z_1 - z_2\|, \ z_1, z_2 \in B(0, K + 2) \cap U. \tag{2.147}$$

Lemma 2.1 implies that there exist $\bar{\alpha}, \gamma, \Delta, \in (0, 1)$ such that the following property holds:

(a) For each $x \in B(0, K_1) \cap C$ satisfying

$$f(x) \geq \inf(f, C) + \epsilon/4,$$

each $l \in E(x, \Delta, \Delta)$, and each $\beta \in (0, \bar{\alpha}]$, the inequality

$$f(x) - f(x + \beta l) \geq \beta \gamma$$

holds.

Assume that sequences $\{x_i\}_{i=0}^{\infty} \subset C$ and $\{l_i\}_{i=0}^{\infty} \subset X$ satisfy

$$\liminf_{i \to \infty} \|x_i\| < K, \tag{2.148}$$

and for each integer $i \geq 0$,

$$l_i \in E(x_i, \Delta, \Delta) \tag{2.149}$$

and

$$x_{i+1} = x_i + \alpha_i l_i. \tag{2.150}$$

Since

$$\lim_{i \to \infty} \alpha_i = 0,$$

there exists a natural number N_1 such that for each integer $i \geq N_1$,

$$\alpha_i < \min\{\bar{\alpha}, (8c_* L_0)^{-1} \epsilon\}. \tag{2.151}$$

In view (2.148), there exists an integer $N_2 > N_1 + 2$ such that

$$\|x_{N_2}\| < K. \tag{2.152}$$

Since

$$\sum_{i=0}^{\infty} \alpha_i = \infty,$$

there exists an integer $N_0 > N_2 + 2$ such that

$$\sum_{i=N_2+1}^{N_0-1} \alpha_i > \gamma^{-1}(K_0 - \inf(f, C)). \tag{2.153}$$

In order to complete the proof of the theorem, it is sufficient to show that

$$f(x_i) \leq \inf(f, C) + \epsilon$$

for all integers $i \geq N_0$.

First we show that there exists an integer $j \in \{N_2, \ldots, N_0\}$ such that

$$f(x_j) \leq \inf(f, C) + \epsilon/4.$$

Assume the contrary. Then

$$f(x_i) > \inf(f, C) + \epsilon/4, \ i = N_2, \ldots, N_0. \tag{2.154}$$

Let

$$i \in \{N_2, \ldots, N_0\}. \tag{2.155}$$

Assume that

$$\|x_i\| \leq K_1. \tag{2.156}$$

(Note that in view of (2.144), (2.146), and (2.152), Eq. (2.156) is true for $i = N_2$.) Property (a) and equations (2.149)–(2.151), (2.154), and (2.155) imply that

$$f(x_i) - f(x_{i+1}) \geq \alpha_i \gamma.$$

Thus we have shown that the following property holds:
(b) If $i \in \{N_2, \ldots, N_0\}$ and $\|x_i\| \leq K_1$, then

$$f(x_i) - f(x_{i+1}) \geq \alpha_i \gamma.$$

We show that

$$\|x_i\| \leq K_1, \ i = N_2, \ldots, N_0.$$

Assume the contrary. Then in view of (2.152), there exists an integer $p \in (N_2, N_0]$ such that

$$\|x_p\| > K_1, \tag{2.157}$$

$$\|x_i\| \leq K_1, \ i = N_2, \ldots, p-1. \tag{2.158}$$

Property (b) and Eqs. (2.157) and (2.158) imply that for each integer $i \in \{N_2, \ldots, p-1\}$,

$$f(x_{i+1}) \leq f(x_i) - \alpha_i \gamma \leq f(x_i),$$

for all $i = N_2, \ldots, p$,

$$f(x_i) \leq f(x_{N_2}) \leq K_0$$

and

$$f(x_p) \leq K_0, \ \|x_p\| \leq K_1.$$

This contradicts (2.157). The contradiction we have reached proves that

$$\|x_i\| \leq K_1, \ i = N_2, \ldots, N_0. \tag{2.159}$$

Property (b) and (2.159) imply that for all integers $i = N_2, \ldots, N_0 - 1$,

$$f(x_i) - f(x_{i+1}) \geq \alpha_i \gamma. \tag{2.160}$$

It follows from (2.144), (2.152), and (2.160) that

$$K_0 - \inf(f, C) \geq f(x_{N_2}) - f(x_{N_0})$$

$$= \sum_{i=N_2}^{N_0-1} (f(x_i) - f(x_{i+1})) \geq \gamma \sum_{i=N_2}^{N_0-1} \alpha_i$$

and

$$\sum_{i=N_2}^{N_0-1} \alpha_i \leq (K_0 - \inf(f, C))\gamma^{-1}.$$

This contradicts (2.153). The contradiction we have reached proves that there exists

$$j \in \{N_2, \ldots, N_0\} \tag{2.161}$$

such that

$$f(x_j) \leq \inf(f, C) + \epsilon/4. \tag{2.162}$$

We show that for each integer $i \geq j$,

$$f(x_i) \leq \inf(f, C) + \epsilon.$$

Assume the contrary. Then there exists an integer $k > j$ such that

$$f(x_k) > \inf(f, C) + \epsilon. \tag{2.163}$$

We may assume without loss of generality that

$$f(x_i) \leq \inf(f, C) + \epsilon, \quad i = j, \ldots, k - 1. \tag{2.164}$$

By (2.142), (2.143), and (2.164),

$$\|x_{k-1}\| \leq K - 2. \tag{2.165}$$

There are two cases:

$$f(x_{k-1}) > \inf(f, C) + \epsilon/4; \tag{2.166}$$

$$f(x_i) \leq \inf(f, C) + \epsilon/4. \tag{2.167}$$

Assume that (2.166) holds. Property (a), the inequality $k - 1 \geq j \geq N_2$, (2.150), (2.151), (2.165), and (2.166) imply that

$$f(x_{k-1}) - f(x_k) = f(x_{k-1}) - f(x_{k-1} + \alpha_{k-1}l_{k-1}) \geq \alpha_{k-1}\gamma$$

and

$$f(x_k) \leq f(x_{k-1}) \leq \inf(f, C) + \epsilon.$$

This contradicts (2.163). The contradiction we have reached proves that (2.167) holds. By (2.149)–(2.151) and the inequality $k - 1 \geq N_2$,

$$\|x_k - x_{k-1}\| \le \alpha_{k-1}\|l_{k-1}\| \le (c_* + 1)\alpha_{k-1} \le 2c_*\alpha_{k-1} \le 1. \tag{2.168}$$

In view of (2.165) and (2.168),

$$\|x_k\| \le K - 1.$$

It follows from (2.147), (2.165), (2.168), and the equation above and the inequality $k - 1 \ge N_2$ that

$$|f(x_k) - f(x_{k-1})| \le L_0\|x_k - x_{k-1}\| \le 2L_0c_*\alpha_{k-1} \le \epsilon/4.$$

Together with (2.167), this implies that

$$f(x_k) \le f(x_{k-1}) + \epsilon/4 \le \inf(f, C) + \epsilon/2.$$

This contradicts (2.163). The contradiction we have reached proves that

$$f(x_i) \le \inf(f, C) + \epsilon$$

for all integers $i \ge j$. Theorem 2.8 is proved.

2.11 Proof of Theorem 2.9

Note that

$$\lim_{\|x\| \to \infty, \, x \in U} f(x) = \infty. \tag{2.169}$$

In view of (2.169), we may assume without loss of generality that $\epsilon < 1$, $K > 2$ and that

$$\{x \in U : f(x) \le \inf(f, C) + 4\} \subset B(0, K - 2). \tag{2.170}$$

Fix

$$K_0 > \sup\{f(x) : x \in B(0, K + 1) \cap C\}. \tag{2.171}$$

Set

$$E_0 = \{x \in U : f(x) \le K_0 + 1\}. \tag{2.172}$$

Clearly, E_0 is a bounded set. Fix

$$K_1 > \sup\{\|x\| : x \in E_0\}. \tag{2.173}$$

There exists $L_0 \ge 1$ such that

$$|f(z_1) - f(z_2)| \le L_0\|z_1 - z_2\|, \; z_1, z_2 \in B(0, K + 2) \cap U. \tag{2.174}$$

Lemma 2.1 implies that there exist $\alpha, \gamma, \Delta, \in (0, 1)$ such that the following property holds:

(a) For each $x \in B(0, K_1) \cap C$ satisfying

$$f(x) \geq \inf(f, C) + \epsilon/4,$$

each $l \in E(x, \Delta, \Lambda)$, and each $\beta \in (0, \alpha]$, the inequality

$$f(x) - f(x + \beta l) \geq \beta \gamma$$

holds.

We may assume without loss of generality that

$$\alpha \leq (8c_*(L_0 + 1))^{-1}\epsilon. \tag{2.175}$$

Assume that

$$\{\alpha_i\}_{i=0}^{\infty} \subset (0, \alpha], \quad \sum_{i=0}^{\infty} \alpha_i = \infty. \tag{2.176}$$

In view of (2.176), there exists a natural number $N_0 > 4$ such that

$$\sum_{i=2}^{N_0-1} \alpha_i > \gamma^{-1}(K_0 - \inf(f, C)). \tag{2.177}$$

Assume that $\{x_i\}_{i=0}^{\infty} \subset C$, $\{l_i\}_{i=0}^{\infty} \subset X$,

$$\|x_0\| \leq K \tag{2.178}$$

and that for each integer $i \geq 0$,

$$l_i \in E(x_i, \Delta, \Lambda) \tag{2.179}$$

and

$$x_{i+1} = x_i + \alpha_i l_i. \tag{2.180}$$

In order to complete the proof of the theorem, it is sufficient to show the inequality

$$f(x_i) \leq \inf(f, C) + \epsilon$$

holds for all integers $i \geq N_0$. Property (a) and Eqs. (2.176) and (2.179) imply that the following property holds:

(b) If an integer $i \geq 0$, $\|x_i\| \leq K_1$ and if

$$f(x_i) \geq \inf(f, C) + \epsilon/4,$$

then

$$f(x_i) - f(x_{i+1}) \geq \alpha_i \gamma.$$

First we show that there exists an integer $i \in \{0, \ldots, N_0\}$ such that

$$f(x_i) \leq \inf(f, C) + \epsilon/4.$$

Assume the contrary. Then

$$f(x_i) > \inf(f, C) + \epsilon/4, \ i = 0, \ldots, N_0. \tag{2.181}$$

We show that

$$\|x_i\| \leq K_1, \ i = 0, \ldots, N_0.$$

Assume the contrary. Then there exists an integer

$$p \in [0, N_0] \tag{2.182}$$

such that

$$\|x_p\| > K_1. \tag{2.183}$$

By (2.178), (2.182), and (2.183), $p > 0$. We may assume without loss of generality that

$$\|x_i\| \leq K_1, \ i = 0, \ldots, p - 1. \tag{2.184}$$

Property (b) and equations (2.171), (2.178), (2.181), (2.182), and (2.184) imply that for each $i \in \{0, \ldots, p - 1\}$,

$$f(x_{i+1}) \leq f(x_i)$$

and

$$f(x_p) \leq f(x_0) \leq K_0. \tag{2.185}$$

By (2.172), (2.173), and (2.185),

$$\|x_p\| \leq K_1.$$

This contradicts (2.183). The contradiction we have reached proves that

$$\|x_i\| \leq K_1, \ i = 0, \ldots, N_0. \tag{2.186}$$

Property (b) and Eqs. (2.181) and (2.186) imply that for all integers $i = 0, \ldots, N_0 - 1$,

$$f(x_i) - f(x_{i+1}) \geq \alpha_i \gamma. \tag{2.187}$$

By (2.185) and (2.187),

$$K_0 - \inf(f, C) \geq f(x_0) - f(x_{N_0})$$

$$= \sum_{i=0}^{N_0-1} (f(x_i) - f(x_{i+1})) \geq \gamma \sum_{i=0}^{N_0-1} \alpha_i$$

and

$$\sum_{i=0}^{N_0-1} \alpha_i \le (K_0 - \inf(f,C))\gamma^{-1}.$$

This contradicts (2.177). The contradiction we have reached proves that there exists

$$j \in \{0, \ldots, N_0\} \tag{2.188}$$

such that

$$f(x_j) \le \inf(f,C) + \epsilon/4. \tag{2.189}$$

We show that for each integer $i \ge j$,

$$f(x_i) \le \inf(f,C) + \epsilon.$$

Assume the contrary. Then there exists an integer

$$p \ge j \tag{2.190}$$

such that

$$f(x_p) > \inf(f,C) + \epsilon. \tag{2.191}$$

In view of (2.189)–(2.191),

$$p > j.$$

We may assume without loss of generality that

$$f(x_i) \le \inf(f,C) + \epsilon, \ i = j, \ldots, p-1. \tag{2.192}$$

By (2.170) and (2.192), for all $i = j, \ldots, p-1$,

$$\|x_i\| \le K - 2. \tag{2.193}$$

There are two cases:

$$f(x_{p-1}) \ge \inf(f,C) + \epsilon/4; \tag{2.194}$$

$$f(x_{p-1}) < \inf(f,C) + \epsilon/4. \tag{2.195}$$

If (2.194) holds, then property (b) and Eqs. (2.176), (2.179), (2.192), and (2.193) imply that

$$f(x_p) \le f(x_{p-1}) \le \inf(f,C) + \epsilon.$$

This contradicts (2.191). The contradiction we have reached proves that (2.195) holds. By (2.175), (2.179), and (2.180),

$$\|x_p - x_{p-1}\| \le \alpha_{p-1}\|l_{p-1}\| \le (c_* + 1)\alpha_{p-1} \le 2c_*\alpha_{p-1} \le 1. \tag{2.196}$$

In view of (2.193) and (2.196),

$$\|x_p\| \le K.$$

It follows from (2.174)–(2.176), (2.193) and (2.195), (2.196) and the equation above that

$$|f(x_p) - f(x_{p-1})| \leq 2L_0 c_* \alpha_{p-1} \leq 2c_* L_0 \alpha \leq \epsilon/4.$$

Together with (2.195), this implies that

$$f(x_p) \leq f(x_{p-1}) + \epsilon/4 \leq \inf(f, C) + \epsilon/2.$$

This contradicts (2.191). The contradiction we have reached proves that

$$f(x_i) \leq \inf(f, C) + \epsilon$$

for all integers $i \geq j$. Theorem 2.9 is proved.

2.12 Proof of Theorem 2.10

Note that

$$\lim_{\|x\| \to \infty, \, x \in U} f(x) = \infty. \tag{2.197}$$

In view of (2.197), we may assume without loss of generality that $\epsilon < 1$, $K > 2$ and that

$$\{x \in U : f(x) \leq \inf(f, C) + 4\} \subset B(0, K - 2). \tag{2.198}$$

Fix

$$K_0 > \sup\{f(x) : x \in B(0, K + 1) \cap U\}. \tag{2.199}$$

Set

$$E_0 = \{x \in U : f(x) \leq K_0 + 1\}. \tag{2.200}$$

Clearly, E_0 is a bounded set. Fix

$$K_1 > \sup\{\|x\| : x \in E_0\}. \tag{2.201}$$

There exists $L_0 \geq 1$ such that

$$|f(z_1) - f(z_2)| \leq L_0 \|z_1 - z_2\|, \quad z_1, z_2 \in B(0, K_1 + 2) \cap U. \tag{2.202}$$

Lemma 2.1 implies that there exist $\alpha, \gamma, \Delta, \in (0, 1)$ such that the following property holds:

(a) For each $x \in B(0, K_1 + 1) \cap C$ satisfying

$$f(x) \geq \inf(f, C) + \epsilon/4,$$

each $l \in E(x, \Delta, \Delta)$, and each $\beta \in (0, \alpha]$, the inequality

$$f(x) - f(x + \beta l) \geq \beta \gamma$$

holds.

Set

$$\beta_0 = \min\{\alpha, \ (4c_*(L_0 + 1))^{-1}\epsilon\}. \tag{2.203}$$

Let

$$\beta_1 \in (0, \beta_0]. \tag{2.204}$$

Fix an integer

$$N_0 > (K_0 - \inf(f, C))(\beta_1 \gamma)^{-1}. \tag{2.205}$$

Assume that $n \geq N_0$ is an integer, $\{x_i\}_{i=0}^n \subset C$, $\{l_i\}_{i=0}^{n-1} \subset X$,

$$\|x_0\| \leq K \tag{2.206}$$

$$\{\alpha_i\}_{i=0}^{n-1} \subset [\beta_1, \beta_0] \tag{2.207}$$

and that for each integer $i \in \{0, \ldots, n-1\}$,

$$l_i \in E(x_i, \Delta, \Delta) \tag{2.208}$$

and

$$x_{i+1} = x_i + \alpha_i l_i. \tag{2.209}$$

In order to complete the proof of the theorem, it is sufficient to show that

$$f(x_i) \leq \inf(f, C) + \epsilon$$

for all integers $i \in [N_0, n]$. First we show that there exists an integer $j \in \{0, \ldots, N_0\}$ such that

$$f(x_j) \leq \inf(f, C) + \epsilon 4.$$

Assume the contrary. Then

$$f(x_i) > \inf(f, C) + \epsilon/4, \ i = 0, \ldots, N_0. \tag{2.210}$$

Assume that $i \in \{0, \ldots, N_0\}$ and

$$\|x_i\| \leq K_1. \tag{2.211}$$

Property (a) and Eqs. (2.203), (2.207)–(2.211) imply that

$$f(x_i) - f(x_{i+1}) = f(x_i) - f(x_i + \alpha_i l_i) \geq \alpha_i \gamma \geq \beta_1 \gamma.$$

Thus we have shown that the following property holds:
(b) If $i \in \{0, \ldots, N_0\}$ and $\|x_i\| \leq K_1$, then

$$f(x_i) - f(x_{i+1}) \geq \beta_1 \gamma.$$

We show that
$$\|x_i\| \leq K_1, \ i = 0, \ldots, N_0.$$

Assume the contrary. Then there exists an integer

$$k \in [0, N_0] \tag{2.212}$$

such that

$$\|x_k\| > K_1. \tag{2.213}$$

By (2.206), $k > 0$. In view of (2.209), we may assume without loss of generality that

$$\|x_i\| \leq K_1, \ i = 0, \ldots, k - 1. \tag{2.214}$$

Property (b) and Eqs. (2.203), (2.207)–(2.210), (2.212), and (2.214) imply that for each $i \in \{0, \ldots, k - 1\}$,

$$f(x_{i+1}) \leq f(x_i).$$

Together with (2.200) and (2.206), this implies that

$$\|x_k\| \leq K_1.$$

This contradicts (2.213). The contradiction we have reached proves that

$$\|x_i\| \leq K_1, \ i = 0, \ldots, N_0. \tag{2.215}$$

Property (b) and (2.215) imply that for all integers $i = 0, \ldots, N_0 - 1$,

$$f(x_i) - f(x_{i+1}) \geq \beta_1 \gamma. \tag{2.216}$$

By (2.199), (2.206), and (2.216),

$$K_0 - \inf(f, C) \geq f(x_0) - f(x_{N_0})$$

$$= \sum_{i=0}^{N_0-1} (f(x_i) - f(x_{i+1})) \geq \gamma \beta_1 N_0$$

and

$$N_0 \leq (\beta_1 \gamma)^{-1} (K_0 - \inf(f, C)).$$

This contradicts (2.205). The contradiction we have reached proves that there exists

$$j \in \{0, \ldots, N_0\} \tag{2.217}$$

such that

$$f(x_j) \leq \inf(f, C) + \epsilon/4. \tag{2.218}$$

We show that for each integer $i \in \{j, \ldots, n\}$,

$$f(x_i) \leq \inf(f, C) + \epsilon.$$

Assume the contrary. Then there exists an integer

$$p \in \{j, \ldots, n\} \tag{2.219}$$

such that

$$f(x_p) > \inf(f, C) + \epsilon. \tag{2.220}$$

In view of (2.218)–(2.220),

$$p > j.$$

We may assume without loss of generality that

$$f(x_i) \leq \inf(f, C) + \epsilon, \ i = j, \ldots, p - 1$$

and in particular

$$f(x_{p-1}) \leq \inf(f, C) + \epsilon. \tag{2.221}$$

By (2.219) and (2.221),

$$\|x_{p-1}\| \leq K - 2. \tag{2.222}$$

There are two cases:

$$f(x_{p-1}) > \inf(f, C) + \epsilon/4; \tag{2.223}$$

$$f(x_{p-1}) \leq \inf(f, C) + \epsilon/4. \tag{2.224}$$

By (2.207)–(2.209),

$$\|x_p - x_{p.-1}\| \leq \alpha_{p-1}\|l_{p-1}\| \leq (c_* + 1)\beta_0 \leq 2c_*\beta_0. \tag{2.225}$$

In view of (2.202), (2.206), (2.222), and (2.225),

$$\|x_p\| \leq K,$$

$$|f(x_p) - f(x_{p-1})| \leq L_0\|x_p - x_{p-1}\| \leq 2L_0c_*\beta \leq \epsilon/2. \tag{2.226}$$

Assume that (2.223) holds. Then property (a) and equations (2.199)–(2.201), (2.203), (2.207)–(2.209), and (2.221)–(2.223) imply that

$$f(x_{p-1}) - f(x_p) \geq \alpha_{p-1}\gamma$$

and

$$f(x_p) \leq f(x_{p-1}) \leq \inf(f, C) + \epsilon.$$

This contradicts (2.220). The contradiction we have reached proves that (2.224) holds. By (2.224) and (2.226),

$$f(x_p) \leq f(x_{p-1}) + \epsilon/2 \leq \inf(f, C) + \epsilon.$$

This contradicts (2.220). The contradiction we have reached proves that

$$f(x_i) \leq \inf(f, C) + \epsilon$$

for all integers $i \in \{j, \ldots, n\}$. Theorem 2.10 is proved.

2.13 A Convergence Result with Estimations

Assume that $(X, \| \cdot \|)$ is a Banach space, $U \subset X$ is an open convex set,

$$C \subset U$$

is a nonempty closed and convex set, and $f : U \to R^1$ is a convex function that is Frechet differentiable at any point $x \in U$.

In this section, we continue to consider the minimization problem

$$f(x) \to \min, \ x \in C.$$

In the previous sections, we showed that for a given $\epsilon > 0$, there exist a sufficiently small error $\delta, \Delta > 0$ such that after a certain number of iterations we obtain a point $x_s \in K$ satisfying $f(x_s) \leq \inf(f, C) + \epsilon$. Of course, it is interesting to obtain an explicit estimation for δ, Δ. It is done in the theorem below under some additional assumptions on the objective function f. Namely, we assume that its Frechet derivative is Holder continuous on bounded sets and that the corresponding constants are known. In this case, we obtain an explicit dependence δ, Δ and a number of iterations on ϵ. This dependence allows us easily to solve an inverse problem: if we know δ, Δ what ϵ can be obtained?

Let $K \geq 1$, $\lambda \in (0, 1]$ and $c_* \geq 1$ be fixed.
Assume that

$$\inf(f, C) = \inf(f, C \cap B(0, K)), \tag{2.227}$$

$$K_0 \geq \sup\{|f(x)| : \ x \in C \cap B(0, K)\} + 1, \tag{2.228}$$

$$K_1 > \sup\{\|x\| : \ x \in U, \ |f(x)| \leq K_0 + 1\}, \tag{2.229}$$

$L_0 \geq 1$, $\rho \in (0, 1]$ and that

$$|f(z_1) - f(z_2)| \leq L_0 \|z_1 - z_2\|, \ z_1, z_2 \in B(0, K_1 + 1) \cap U, \tag{2.230}$$

$$\|f'(z_1) - f'(z_2)\|_* \leq L_0 \|z_1 - z_2\|^\rho, \ z_1, z_2 \in B(0, K_1 + 1) \cap C. \tag{2.231}$$

Let $\epsilon \in (0, 1]$,

$$\Delta = (64K_1)^{-1} \epsilon \lambda, \tag{2.232}$$

$$\gamma = (64K_1)^{-1} \epsilon \lambda, \tag{2.233}$$

$$\delta = (64K_1)^{-1} \epsilon \lambda (2c_* + L_0)^{-1}, \tag{2.234}$$

$$\alpha = \min\{((2^6 K_1 c_*^2 (L_0 + 1))^{-1} \epsilon \lambda)^{1/\rho}, \ (8c_* L_0)^{-1} \epsilon\}, \tag{2.235}$$

$$\gamma = (64K_1)^{-1}\epsilon\lambda. \tag{2.236}$$

Recall (see (2.1) and (2.2)) that for each $x \in C$ and each pair of numbers $\delta, \Delta \in (0,1]$, we denote by $E(x,\delta,\Delta)$ the set of all

$$l \in C - x \tag{2.237}$$

for which there exist $g \in X^*$ such that

$$\|g - f'(x)\|_* \leq \delta \tag{2.238}$$

and

$$\xi \in B(0,c_*) \cap B(l,\delta) \tag{2.239}$$

such that

$$\langle g, \xi \rangle \leq \lambda \inf\{\langle g, \eta \rangle : \eta \in B(0,c_*) \cap (C - x)\} + \Delta. \tag{2.240}$$

Theorem 2.11. *Assume that*

$$\alpha \in (0,\bar{\alpha}],$$

an integer N_0 satisfies

$$N_0 > 2K_0\bar{\alpha}^{-1}\gamma^{-1}, \tag{2.241}$$

$n \geq N_0$ is an integer,

$$\{x_i\}_{i=0}^n \subset C, \tag{2.242}$$

$\{l_i\}_{i=0}^{n-1} \subset X,$

$$\alpha_i \in [\bar{\alpha},\alpha], \; i = 0,\ldots,n-1, \tag{2.243}$$

$$\|x_0\| \leq K, \tag{2.244}$$

and that for each integer $i \in \{0,\ldots,n-1\}$,

$$l_i \in E(x_i,\delta,\Delta) \tag{2.245}$$

and

$$x_{i+1} = x_i + \alpha_i l_i. \tag{2.246}$$

Then the inequality

$$f(x_i) \leq \inf(f,C) + \epsilon$$

holds for all integers $i \in \{N_0,\ldots,n\}$, and there exists

$$j \in \{0,\ldots,N_0 - 1\}$$

such that

$$f(x_j) - f(x_{j+1}) < \bar{\alpha}\gamma.$$

Moreover, if

$$j \in \{0, \ldots, N_0 - 1\},$$

$$f(x_j) - f(x_{j+1}) < \bar{\alpha}\gamma,$$

and for each integer $i \in \{0, \ldots, j\} \setminus \{j\}$, the inequality

$$f(x_j) - f(x_{i+1}) \geq \bar{\alpha}\gamma$$

is true, then

$$f(x_j) \leq \inf(f, C) + \epsilon.$$

Clearly, N_0 is a number of iterations needed in order to obtain a point $x_j \in C$ satisfying $f(x_j) \leq \inf(f, C) + \epsilon$. Theorem 2.11 answers the question how we can find this j. According to the theorem, j is the number of the first iteration for which

$$f(x_j) - f(x_{j+1}) < \bar{\alpha}\gamma.$$

Theorem 2.11 is proved in Sect. 2.15. Its proof is based on a lemma that is proved in the next section.

2.14 An Auxiliary Result

Lemma 2.12. *Let*

$$x \in B(0, K_1) \cap C \tag{2.247}$$

satisfy

$$f(x) \geq \inf(f, C) + \epsilon/4, \tag{2.248}$$

$$l \in E(x, \delta, \Delta), \tag{2.249}$$

$$\beta \in (0, \alpha]. \tag{2.250}$$

Then

$$f(x) - f(x + \beta l) \geq \beta\gamma.$$

Proof. By (2.227), there exists

$$x_0 \in B(0, K) \cap C \tag{2.251}$$

such that

$$f(x_0) \leq \inf(f; C) + 8^{-1}\epsilon. \tag{2.252}$$

In view of (2.248), (2.252), and convexity of f,

$$x \neq x_0$$

and

$$- \epsilon/8 \geq f(x_0) - f(x) \geq \langle f'(x), x - x_0 \rangle. \tag{2.253}$$

By (2.247) and (2.251),

$$\|x - x_0\| \leq 2K_1. \tag{2.254}$$

Set

$$x_1 = x + (2K_1)^{-1}(x_0 - x) = (2K_1)^{-1}x_0 + (1 - (2K_1)^{-1})x \in C \tag{2.255}$$

in view of (2.247) and (2.251) and set

$$\eta = x_1 - x. \tag{2.256}$$

Equations (2.254)–(2.256) imply that

$$\eta \in C - x, \ \|\eta\| \leq 1. \tag{2.257}$$

By (2.253), (2.255), (2.256) and convexity of f,

$$\langle f'(x), \eta \rangle \leq f(x_1) - f(x)$$

$$\leq (2K_1)^{-1}f(x_0) + (1 - (2K_1)^{-1})f(x) - f(x)$$

$$\leq 2K_1)^{-1}(f(x_0) - f(x)) \leq -(16K_1)^{-1}\epsilon. \tag{2.258}$$

It follows from (2.249) (see (2.1), (2.2), and (2.237)–(2.240)) that there exist

$$\xi \in B(l, \delta) \cap B(0, c_*) \tag{2.259}$$

and $g \in X^*$ such that

$$\|g - f'(x)\|_* \leq \delta \tag{2.260}$$

and

$$\langle g, \xi \rangle \leq \lambda \langle g, \eta \rangle + \Delta. \tag{2.261}$$

By the mean value theorem, there exists $\theta \in [0, \beta]$ such that

$$f(x + \beta l) - f(x) = \langle f'(x + \theta l), \beta l \rangle. \tag{2.262}$$

In view of (2.262),

$$f(x + \beta l) - f(x) \leq \beta \langle f'(x), l \rangle + \|f'(x + \theta l) - f'(x)\|_* \beta \|l\|. \tag{2.263}$$

It follows from (2.230), (2.247), and (2.259) that

$$\beta \langle f'(x), l \rangle \leq \beta \langle f'(x), \xi \rangle + \beta \|f'(x)\|_* \|l - \xi\|$$

$$\leq \beta \langle f'(x), \xi \rangle + \beta \delta L_0. \tag{2.264}$$

By (2.250) and (2.259),

$$\|\theta l\| \leq 2c_*\alpha. \tag{2.265}$$

Equations (2.247) and (2.265) imply that

$$\|x + \theta l\| \leq K_1 + 1.$$

It follows from (2.231), (2.237), (2.247), (2.249), (2.265) and the inequality above that

$$\|f'(x + \theta l) - f'(x)\|_* \leq L_0(2c_*\alpha)^p. \tag{2.266}$$

Equations (2.259), (2.260) (2.263), (2.264), and (2.266) imply that

$$f(x + \beta l) - f(x) \leq \beta\langle f'(x), \xi\rangle + L_0\beta\delta + 2L_0\beta c_*(2c_*\alpha)^p. \tag{2.267}$$

By (2.257)–(2.261),

$$\langle f'(x), \xi\rangle \leq \langle g, \xi\rangle + \|f'(x) - g\|_*\|\xi\| \leq \langle g, \xi\rangle + +c_*\delta$$

$$\leq \lambda\langle g, \eta\rangle + \Delta + \delta c_*$$

$$\leq \lambda\langle f'(x), \eta\rangle + \|f'(x) - g\|_*\|\eta\| + \Delta + \delta c_*$$

$$\leq -\lambda(16K_1)^{-1}\epsilon + \Delta + \delta(1 + c_*).$$

Combined with (2.232)–(2.236) and (2.267), this implies that

$$f(x + \beta l) - f(x)$$

$$\leq \beta(-\lambda(16K_1)^{-1}\epsilon + \Delta + \delta(c_* + 1)$$

$$+\delta L_0 + 4c_*^2 L_0\alpha^p) \leq -\beta\gamma.$$

Lemma 2.12 is proved.

2.15 Proof of Theorem 2.11

We show that there exists an integer $j \in \{0, \ldots, N_0\}$ such that

$$f(x_j) \leq \inf(f, C) + \epsilon/4.$$

Assume the contrary. Then

$$f(x_i) > \inf(f, C) + \epsilon/4, \quad i = 0, \ldots, N_0. \tag{2.268}$$

Assume that $i \in \{0, \ldots, N_0 - 1\}$ and that

$$\|x_i\| \leq K_1, \ f(x_i) \leq f(x_0) \leq K_0. \tag{2.269}$$

(In view of (2.229) and (2.244), Eq. (2.269) holds for $i = 0$.) Lemma 2.12 and Eqs. (2.243), (2.245), (2.246), (2.268), and (2.269) imply that

$$f(x_i) - f(x_{i+1}) = f(x_i) - f(x_i + \alpha_i l_i) \geq \alpha_i \gamma \geq \bar{\alpha}\gamma. \tag{2.270}$$

Together with (2.229), this implies that

$$\|x_{i+1}\| \leq K_1.$$

Thus the assumption made for i also holds for $i + 1$. Therefore by induction, we showed that (2.269) holds for all integers $i = 0, \ldots, N_0$ and (2.270) holds for all integers $i = 0, \ldots, N_0 - 1$. By (2.227), (2.228), (2.269), and (2.270),

$$2K_0 > K_0 - \inf(f, C) \geq f(x_0) - f(x_{N_0})$$

$$= \sum_{i=0}^{N_0 - 1} (f(x_i) - f(x_{i+1})) \geq \gamma\bar{\alpha}N_0$$

and

$$N_0 \leq 2K_0(\bar{\alpha}\gamma)^{-1}.$$

This contradicts (2.241). The contradiction we have reached proves that there exists

$$j \in \{0, \ldots, N_0\} \tag{2.271}$$

such that

$$f(x_j) \leq \inf(f, C) + \epsilon/4.$$

Assume that an integer $i \in \{j, \ldots, n\} \setminus \{n\}$ and

$$f(x_i) \leq \inf(f, C) + \epsilon. \tag{2.272}$$

We show that

$$f(x_{i+1}) \leq \inf(f, C) + \epsilon.$$

There are two cases:

$$f(x_i) > \inf(f, C) + \epsilon/4; \tag{2.273}$$

$$f(x_i) \leq \inf(f, C) + \epsilon/4. \tag{2.274}$$

By (2.227)–(2.229) and (2.272),

$$\|x_i\| \leq K_1. \tag{2.275}$$

Assume that (2.273) holds. Lemma 2.12 and Eqs. (2.243), (2.245), (2.246), (2.272), (2.273), and (2.275) imply that

$$f(x_i) - f(x_{i+1}) = f(x_i) - f(x_i + \alpha_i l_i) \geq \alpha_i \gamma$$

and

$$f(x_{i+1}) \leq f(x_i) \leq \inf(f, C) + \epsilon.$$

Assume that (2.274) is true. By (2.243) and (2.246),

$$\|x_i - x_{i+1}\| \leq \alpha_i \|l_i\| \leq \alpha \|l_i\|. \tag{2.276}$$

In view of (2.239) and (2.245), there exists $\xi \in X$ such that

$$\|l_i - \xi\| \leq \delta, \ \|\xi\| \leq c_*. \tag{2.277}$$

It follows from (2.235), (2.276), and (2.277) that

$$\|x_i - x_{i+1}\| \leq \alpha(\|\xi\| + 1) \leq 2\alpha c_* \leq 1. \tag{2.278}$$

Equations (2.275) and (2.278) imply that

$$\|x_{i+1}\| \leq K_1 + 1.$$

By (2.230), (2.275), (2.278),

$$|f(x_i) - f(x_{i+1})| \leq L_0 \|x_i - x_{i+1}\| \leq 2L_0 c_* \alpha. \tag{2.279}$$

In view of (2.235), (2.274) and (2.279),

$$f(x_{i+1}) \leq f(x_i) + 2\alpha c_* L_0 \leq \inf(f, C) + \epsilon/4 + 2\alpha c_* L_0 \leq \inf(f, C) + \epsilon.$$

Therefore by induction, we showed that

$$f(x_i) \leq \inf(f, C) + \epsilon$$

for all integers $i \in \{j, \ldots, n\}$.

Assume that for each $j \in \{0, \ldots, N_0 - 1\}$,

$$f(x_j) - f(x_{j+1}) \geq \bar{\alpha} \gamma.$$

Then in view of (2.227), (2.228), and (2.244),

$$f(x_i) \leq f(x_0) \leq K_0, \ i = 0, \ldots, N_0,$$

$$2K_0 \geq f(x_0) - \inf(f, C) \geq f(x_0) - f(x_{N_0})$$

$$= \sum_{i=0}^{N_0-1} (f(x_i) - f(x_{i+1})) \geq \gamma \bar{\alpha} N_0$$

and

$$N_0 \leq 2K_0(\bar{\alpha}\gamma)^{-1}.$$

This contradicts (2.241). The contradiction we have reached proves that there exists

$$j \in \{0, \ldots, N_0 - 1\}$$

such that

$$f(x_j) - f(x_{j+1}) < \bar{\alpha}\gamma. \tag{2.280}$$

Assume that

$$j \in \{0, \ldots, N_0 - 1\},$$

(2.280) holds, and for every integer $i \in \{0, \ldots, j\} \setminus \{j\}$,

$$f(x_j) - f(x_{i+1}) \geq \bar{\alpha}\gamma.$$

It follows from (2.228), (2.229), and (2.244) that

$$f(x_i) \leq f(x_0) \leq K_0, \ i = 0, \ldots, j$$

and

$$\|x_i\| \leq K_1, \ i = 0, \ldots, j. \tag{2.281}$$

If

$$f(x_j) \geq \inf(f, C) + \epsilon/4,$$

then Lemma 2.12, (2.240), (2.243), (2.245), and (2.281) imply that

$$f(x_j) - f(x_{i+1}) \geq \bar{\alpha}\gamma,$$

which contradicts (2.280). The contradiction we have reached proves that

$$f(x_j) \leq \inf(f, C) + \epsilon/4.$$

This completes the proof of Theorem 2.11.

3

Nonconvex Optimization

In this chapter, we study the algorithms for constrained nonconvex minimization problems in a general Banach space with Frechet differentiable objective functions. Our goal is to obtain a good approximate solution of the problem in the presence of computational errors. It is shown that the algorithm generates a good approximate solution, if the sequence of computational errors is bounded from above by a small constant. We obtain a number of convergence results under different conditions including a theorem with explicit estimations for computational errors.

3.1 Preliminaries

Assume that $(X, \|\cdot\|)$ is a Banach space, $U \subset X$ is an open convex set,

$$C \subset U$$

is a nonempty closed and convex set, and a function $f : U \to R^1$ is Frechet differentiable at any point $x \in U$ and is Lipschitz on all bounded sets in U.

In this section, we consider the minimization problem

$$f(x) \to \min, \ x \in C. \qquad (P)$$

Since the objective function f is nonconvex in the case when $U = X$, iterative gradient-type methods allow us to construct a sequence of iterates $\{x_t\}_{t=0}^{\infty}$ such that $\|f'(x_t)\|_* \to 0$ when $t \to \infty$. It means that actually we search for a solution of the problem

A. J. Zaslavski, *Optimization in Banach Spaces*, SpringerBriefs in Optimization, https://doi.org/10.1007/978-3-031-12644-4_3

$$\|f'(x)\|_* \to \min, \ x \in X.$$

For our constrained problem (P), the situation is more difficult. It is natural to minimize the function $\Lambda : C \to R^1$ that is defined below.

For each $x \in C$, set

$$\Lambda(x) = \inf\{\langle f'(x), h \rangle : \ h \in (C - x) \cap B(0, 1)\}. \tag{3.1}$$

Clearly, $\Lambda(x) \leq 0$ for each $x \in C$. (Note that if $C = U = X$, then $\Lambda(x) = -\|f'(x)\|_*, x \in X$.) Our goal is to find an approximate solution of the problem

$$\Lambda(x) \to max, \ x \in C.$$

More precisely, we wish to find $x \in C$ satisfying $\Lambda(x) \geq -\epsilon$, where ϵ is a sufficiently small positive number. Clearly, if x is a solution of problem (P), then $\Lambda(x) = 0$.

We assume that the following assumption holds.

(A1) For every $M > 0$, the Frechet derivative $f'(\cdot)$ is uniformly continuous on the set $B(0, M) \cap C$.

Let $c_* > 1$ and $\lambda \in (0, 1]$ be fixed. In our study, we use the same algorithms $\mathcal{A}1 - \mathcal{A}3$ that were used in Chap. 2 for the convex minimization problem. We recall the following definition.

For each $x \in C$ and each pair of numbers $\delta, \Delta \in (0, 1]$, denote by $E(x, \delta, \Delta)$ the set of all $l \in C - x$ for which there exist $g \in X^*$ such that

$$\|g - f'(x)\|_* \leq \delta \tag{3.2}$$

and

$$\xi \in B(0, c_*) \cap B(l, \delta) \tag{3.3}$$

such that

$$\langle g, \xi \rangle \leq \lambda \inf\{\langle g, \eta \rangle : \ \eta \in B(0, c_*) \cap (C - x)\} + \Delta. \tag{3.4}$$

The next section contains several auxiliary results that are used in the proofs of our convergence theorems stated in Sect. 3.3.

3.2 Auxiliary Results

Lemma 3.1. *Assume that $x \in C$, $g \in X^*$, and $L > 1$. Then*

$$\inf\{\langle g, h \rangle : \ h \in (C - x) \cap B(0, 1)\}$$

$$\geq \inf\{\langle g, h \rangle : \ h \in (C - x) \cap B(0, L)\}$$

$$\geq L \inf\{\langle f, h \rangle : \ h \in (C - x) \cap B(0, 1)\}.$$

Proof. Clearly,

$$\inf\{\langle g, h\rangle : h \in (C - x) \cap B(0, 1)\}$$

$$\geq \inf\{\langle g, h\rangle : h \in (C - x) \cap B(0, L)\}.$$

Assume that

$$\xi \in (C - x) \cap B(0, L). \tag{3.5}$$

If $\|\xi\| \leq 1$, then

$$\langle g, \xi\rangle \geq \inf\{\langle g, h\rangle : h \in (C - x) \cap B(0, 1)\}.$$

Assume that

$$\|\xi\| > 1.$$

In view of (3.5),

$$x + \xi \in C.$$

By the convexity of C,

$$x + L^{-1}\xi \in C,$$

$$L^{-1}\xi \in C - x, \ \|L^{-1}\xi\| \leq 1,$$

and

$$\langle g, \xi\rangle \geq L \inf\{\langle g, h\rangle : h \in (C - x) \cap B(0, 1)\}.$$

Thus in both cases

$$\langle g, \xi\rangle \geq \inf\{\langle g, h\rangle : h \in (C - x) \cap B(0, 1)\}.$$

Lemma 3.1 is proved.

Proposition 3.2. *The function $\Lambda(\cdot)$ is uniformly continuous on bounded sets.*

Proof. Let $M, \epsilon > 0$. There exist $L > 1$ and $\gamma \in (0, 1)$ such that

$$|f(z_1) - f(z_2)| \leq L\|z_1 - z_2\|, \ z_1, z_2 \in B(0, M + 1) \cap U \tag{3.6}$$

and

$$1 - \gamma < \epsilon/8. \tag{3.7}$$

By (A1), there exists $\delta \in (0, 1)$ such that

$$\delta < \min\{1 - \gamma, \ (8L)^{-1}\epsilon\},$$

$$\|f'(z_1) - f(z_2)\|_* \leq \epsilon/8 \tag{3.8}$$

for each $z_1, z_2 \in B(0, M) \cap C$ satisfying $\|z_1 - z_2\| \leq \delta$.

Assume that

$$x_1, x_2 \in B(0, M) \cap C \text{ and } \|x_1 - x_2\| \leq \delta. \tag{3.9}$$

By (3.1), (3.6), and (3.9),

$$|\Lambda(x_i)| \leq L, \ i = 1, 2.$$

In view of (3.1), there exists

$$h_1 \in (C - x_1) \cap B(0, 1) \tag{3.10}$$

such that

$$|\langle f'(x_1), h_1 \rangle - \Lambda(x_1)| \leq \epsilon/8. \tag{3.11}$$

By (3.10),

$$\|\gamma h_1\| \leq \gamma, \ \gamma h_1 \in C - x_1. \tag{3.12}$$

It follows from (3.8), (3.9), and (3.12) that

$$\gamma h_1 + x_1 - x_2 \in C - x_2 \tag{3.13}$$

and

$$\|\gamma h_1 + x_1 - x_2\| \leq \gamma + \delta < 1. \tag{3.14}$$

Equations (3.1), (3.6)–(3.9), (3.11), (3.13), and (3.14) imply that

$$\Lambda(x_2) \leq \langle f'(x_2), \gamma h_1 + x_2 - x_1 \rangle$$

$$\leq \langle f'(x_1), \gamma h_1 + x_2 - x_1 \rangle + \|f'(x_2) - f'(x_1)\|_*$$

$$\leq \gamma \langle f'(x_1), h_1 \rangle + \|f'(x_1)\|_* \|x_1 - x_2\| + \epsilon/8$$

$$\leq \gamma \Lambda(x_1) + \epsilon/8 + L\delta + \epsilon/8$$

$$\leq \Lambda(x_1) - (1 - \gamma)\Lambda(x_1) + \epsilon/4 + L\delta \leq \Lambda(x_1) + \epsilon.$$

This completes the proof of Proposition 3.2.

Lemma 3.3. *Assume that $x \in C$, $\delta, \Delta \in (0, 1]$, and*

$$l \in E(x, \delta, \Delta). \tag{3.15}$$

Then

$$\langle f'(x_1), l \rangle \leq 2c_*\delta + \Delta + \delta(\|f'(x)\|_* + 1).$$

Proof. By (3.15), there exist $\xi \in X$ and $g \in X^*$ satisfying (3.1)–(3.4). It follows from (3.1)–(3.4) that

$$\langle f'(x_1), l \rangle \leq \|g - f'(x)\|_* \|l\| + \langle g, l \rangle \leq 2c_*\delta + \langle g, l \rangle. \tag{3.16}$$

Equations (3.2)–(3.4) imply that

$$\langle g, l \rangle \leq \langle g, \xi \rangle + \|g\| \|l - \xi\|$$

$$\leq \langle g, \xi \rangle + (\|f'(x)\|_* + 1)\delta \leq (\|f'(x)\|_* + 1)\delta + \Delta. \tag{3.17}$$

By (3.10) and (3.17),

$$\langle f'(x_1), l \rangle \leq 2c_*\delta + \langle g, l \rangle \leq 2c_*\delta + \Delta + \delta(\|f'(x)\|_* + 1).$$

Lemma 3.3 is proved.

Lemma 3.4. *Let $\bar{K} \geq 1$, $\bar{\epsilon} \in (0, 1]$. Then there exist $\alpha, \gamma, \Delta, \delta \in (0, 1)$ such that for each $x \in B(0, \bar{K}) \cap C$ satisfying*

$$\Lambda(x) \leq -\bar{\epsilon},$$

each $l \in E(x, \delta, \Delta)$, and each $\beta \in (0, \alpha]$, the inequality

$$f(x) - f(x + \beta l) \geq \beta\gamma$$

holds.

Proof. Since f is Lipschitz on bounded sets, there exists $L_0 \geq 1$ such that

$$|f(z_1) - f(z_2)| \leq L_0 \|z_1 - z_2\|, \ z_1, z_2 \in B(0, \bar{K} + 1) \cap U. \tag{3.18}$$

Fix a positive number

$$\delta_1 \leq (16c_*)^{-1}\bar{\epsilon}\lambda. \tag{3.19}$$

By assumption (A2), there exists $\delta_2 \in (0, 1)$ such that the following property holds:

(a) For each pair of points $z_1, z_2 \in B(0, \bar{K} + 1) \cap C$ satisfying

$$\|z_1 - z_2\| \leq \delta_2,$$

we have

$$\|f'(z_1) - f'(z_2)\|_* \leq \delta_1.$$

Choose positive numbers

$$\delta \leq \min\{\delta_2,\ 8^{-1}\bar{\epsilon}\lambda(L_0 + 2c_*)^{-1}\}, \tag{3.20}$$

$$\Delta \leq 8^{-1}\bar{\epsilon}\lambda, \tag{3.21}$$

$$\gamma = \bar{\epsilon}\lambda/2, \tag{3.22}$$

and

$$\alpha \leq (2c_*)^{-1}\delta. \tag{3.23}$$

Assume that

$$x \in B(0, \bar{K}) \cap C \tag{3.24}$$

satisfies

$$\Lambda(x) \leq -\bar{\epsilon}, \tag{3.25}$$

$$l \in E(x, \delta, \Delta), \tag{3.26}$$

and

$$\beta \in (0, \alpha]. \tag{3.27}$$

It follows from (3.2)–(3.4) and (3.26) that there exist

$$\xi \in B(0, c_*) \cap B(l, \delta) \tag{3.28}$$

and $g \in X^*$ such that

$$\|g - f'(x)\|_* \leq \delta \tag{3.29}$$

and

$$\langle g, \xi \rangle \leq \lambda \inf\{\langle g, \eta \rangle : \eta \in B(0, c_*) \cap (C - x)\} + \Delta. \tag{3.30}$$

Lemma 3.1 and Eqs. (3.1), (3.18), (3.24), (3.26), and (3.28)–(3.30) imply that

$$\langle f'(x), l \rangle \leq \langle f'(x), \xi \rangle + \|f'(x)\|_* \|\xi - l\|$$

$$\leq \langle f'(x), \xi \rangle + L_0\delta$$

$$\leq \langle g, \xi \rangle + \|f'(x) - g\|_* \|\xi\| + L_0\delta$$

$$\leq \langle g, \xi \rangle + \delta c_* + L_0\delta$$

$$\leq \delta(c_* + L_0) + \lambda \inf\{\langle g, \eta \rangle : \eta \in B(0, c_*) \cap (C - x)\} + \Delta$$

$$\leq \delta(c_* + L_0) + \Delta + \lambda(\inf\{\langle f'(x), \eta \rangle : \eta \in B(0, c_*) \cap (C - x)\} + \delta c_*)$$

$$\leq \delta(2c_* + L_0) + \Delta + \lambda \inf\{\langle f'(x), \eta \rangle : \eta \in B(0, c_*) \cap (C - x)\}$$

$$\leq \Delta + \delta(2c_* + L_0) + \lambda\Lambda(x) \leq \Delta + \delta(2c_* + L_0) - \lambda\bar{\epsilon}. \tag{3.31}$$

By the mean value theorem, there exists $\theta \in [0, \beta]$ such that

$$f(x + \beta l) - f(x) = \langle f'(x + \theta l), \beta l \rangle. \tag{3.32}$$

Equations (3.23), (3.26), and (3.27) imply that

$$0 \leq \beta \leq \alpha < 1, l \in C - x. \tag{3.33}$$

In view of (3.33),

$$x + \theta l, \; x + \beta l \in C. \tag{3.34}$$

By (3.28),

$$\|l\| \leq \|\xi\| + 1 \leq c_* + 1 \leq 2c_*.$$

By (3.20), (3.23), (3.33), and (3.34),

$$\|\theta l\| \leq \|\beta l\| \leq 2c_*\alpha \leq \delta \leq \delta_2. \tag{3.35}$$

It follows from (3.23), (3.24), and (3.35) that

$$\|x + \theta l\|, \; \|x + \beta l\| \leq \|x\| + \beta\|l\| \leq \bar{K} + 2c_*\alpha \leq \bar{K} + 1. \tag{3.36}$$

Equations (3.18), (3.24), (3.35), and (3.36) imply that

$$|f(x + \theta l) - f(x)| \leq L_0\|\beta l\| \leq 2\alpha L_0 c_*. \tag{3.37}$$

By (3.32),

$$f(x + \beta l) - f(x) \leq \beta\langle f'(x), l \rangle + \|f'(x + \theta l) - f'(x)\|_*\beta\|l\|. \tag{3.38}$$

Property (a) and Eqs. (3.24), (3.34)–(3.36) imply that

$$\|f'(x + \theta l) - f'(x)\|_* \leq \delta_1. \tag{3.39}$$

Equations (3.38) and (3.39) imply that

$$f(x + \beta l) - f(x) \leq \beta\langle f'(x), l \rangle + 2c_*\beta\delta_1.$$

It follows from the equation above, (3.19)–(3.23) and (3.31), that

$$f(x + \beta l) - f(x)$$

$$\leq 2\beta c_*\delta_1 + \beta(\Delta + \delta(2c_* + L_0) - \lambda\bar{\epsilon})$$

$$= \beta(-\lambda\bar{\epsilon} + \Delta + \delta(2c_* + L_0) + 2\delta_1 c_*)$$

$$\leq -\beta\lambda\bar{\epsilon}/2 \leq -\beta\gamma.$$

Lemma 3.4 is proved.

3.3 Convergence Results

We continue to use the notation, definitions, and assumptions introduced in Sect. 3.1. In this section, we state our results that describe the behavior of algorithms $\mathcal{A}1 - \mathcal{A}3$.

We assume that $c_* \geq 1$ and $\lambda \in (0,1]$ are fixed. In our first result related to algorithm $\mathcal{A}1$, we assume that the sequence of iterates is bounded. It is proved in Sect. 3.4.

Theorem 3.5. *Assume that a sequence* $\{x_i\}_{i=0}^{\infty} \subset C$ *is bounded, and for each integer* $i \geq 0$,

$$l_i \in E(x_i, 0, 0), \ \alpha_i \in [0,1]$$

$$f(x_i + \alpha_i l_i) \leq f(x_i + \beta l_i), \ \beta \in [0,1]$$

and

$$x_{i+1} = x_i + \alpha_i l_i.$$

Then $\liminf_{i \to \infty} \Lambda(x_i) \geq 0$.

In our second result related to algorithm $\mathcal{A}2$, we assume that the sequence of iterates is bounded. It is proved in Sect. 3.4.

Theorem 3.6. *Assume that* $\{\alpha_t\}_{t=0}^{\infty} \subset (0,1)$ *satisfies*

$$\lim_{i \to \infty} \alpha_i = 0, \ \sum_{i=0}^{\infty} \alpha_i = \infty,$$

a sequence $\{x_i\}_{i=0}^{\infty} \subset C$ *is bounded, and that for each integer* $i \geq 0$,

$$l_i \in E(x_i, 0, 0)$$

and if

$$f(x_i + \alpha_i l_i) < f(x_i),$$

then

$$x_{i+1} = x_i + \alpha_i l_i;$$

otherwise, $x_{i+1} = x_i$. *Then* $\limsup_{i \to \infty} \Lambda(x_i) \geq 0$.

In our third result related to algorithm $\mathcal{A}1$, we assume that the objective function f satisfies a growth condition. It is proved in Sect. 3.4.

Theorem 3.7. *Assume that*

$$\lim_{\|x\| \to \infty, \ x \in U} f(x) = \infty$$

and $K, \epsilon > 0$. *Then there exist a natural number* N_0 *and* $\delta > 0$ *such that for each integer* $n \geq N_0$ *and each sequence* $\{x_i\}_{i=0}^{n} \subset C$ *such that*

$$\|x_0\| \le K$$

and for each integer $i \in \{0, \dots, n-1\}$,

$$l_i \in E(x_i, \delta, \delta), \ \alpha_i \in (0, 1],$$

$$f(x_i + \alpha_i l_i) \le f(x_i + \beta l_i), \ \beta \in [0, 1],$$

and

$$x_{i+1} = x_i + \alpha_i l_i.,$$

the inequality

$$Card(\{i \in \{0, \dots, n-1\} : \Lambda(x_i) \le -\epsilon\}) \le N_0$$

holds.

In the following result related to algorithm $\mathcal{A}2$, we assume that the objective function f satisfies a growth condition. It is proved in Sect. 3.4.

Theorem 3.8. *Assume that*

$$\lim_{\|x\| \to \infty, \ x \in U} f(x) = \infty$$

and $K, \epsilon > 0$. *Then there exists* $\delta \in (0, 1)$ *such that for each* $\{\alpha_t\}_{t=0}^{\infty} \subset (0, 1)$ *that satisfies*

$$\lim_{i \to \infty} \alpha_i = 0, \ \sum_{i=0}^{\infty} \alpha_i = \infty,$$

there exists a natural number N for which the following assertion holds.
For each sequence $\{x_i\}_{i=0}^{N} \subset C$ *and each sequence* $\{l_i\}_{i=0}^{N-1} \subset X$ *such that*

$$\|x_0\| \le K$$

and for each integer $i \in \{0, \dots, N-1\}$,

$$l_i \in E(x_i, \delta, \delta),$$

and if

$$f(x_i + \alpha_i l_i) < f(x_i),$$

then

$$x_{i+1} = x_i + \alpha_i l_i;$$

otherwise, $x_{i+1} = x_i$ *the inequality*

$$\max\{\Lambda(x_i) : \ i = 0, \dots, N-1\} \ge -\epsilon$$

is true.

In our following two results related to algorithm $\mathcal{A}1$, we assume that the objective function f satisfies a growth condition. The first of them is proved in Sect. 3.4, while the second one in Sect. 3.5.

Theorem 3.9. *Assume that*

$$\lim_{\|x\|\to\infty,\ x\in U} f(x) = \infty$$

and $K, \epsilon > 0$. Then there exist $\delta, \delta_0 > 0$ and a natural number N_0 such that for each sequence $\{x_i\}_{i=0}^{N_0} \subset C$ and each sequence $\{l_i\}_{i=0}^{N_0-1} \subset X$ such that

$$\|x_0\| \leq K$$

and for each integer $i \in \{0, \ldots, N_0 - 1\}$,

$$l_i \in E(x_i, \delta, \delta), \ \alpha_i \in [0, 1],$$

$$f(x_i + \alpha_i l_i) \leq f(x_i + \beta l_i) + \delta_0, \ \beta \in [0, 1],$$

and

$$x_{i+1} = x_i + \alpha_i l_i,$$

the inequality

$$\max\{\Lambda(x_i):\ i = 0, \ldots, N_0 - 1\} \geq -\epsilon$$

is true.

Theorem 3.10. *Assume that*

$$\lim_{\|x\|\to\infty,\ x\in U} f(x) = \infty$$

and $K, \epsilon > 0$. Then there exist $\delta, \delta_0 > 0$ and a natural number N_0 such that for each sequence $\{x_i\}_{i=0}^{\infty} \subset C$ and each sequence $\{l_i\}_{i=0}^{\infty} \subset X$ such that

$$\|x_0\| \leq K$$

and for each integer $i \geq 0$,

$$l_i \in E(x_i, \delta, \delta),$$

$$x_{i+1} \in \{x_i + \beta l_i:\ \beta \in [0, 1]\},$$

and

$$f(x_{i+1}) \leq f(x_i + \beta l_i) + \delta_0, \ \beta \in [0, 1],$$

$$f(x_{i+1}) \leq f(x_i),$$

the inequality

$$Card(\{i \in \{0, 1, \ldots, \} : \Lambda(x_i) \leq -\epsilon\}) \leq N_0$$

is true.

In our next result related to algorithm $\mathcal{A}1$, we assume that the sequence of iterates is bounded. It is proved in Sect. 3.5.

Theorem 3.11. *Assume that* $\{\delta_i\}_{i=0}^\infty \subset (0,1]$, $\{\epsilon_i\}_{i=0}^\infty \subset (0,1]$ *satisfy*

$$\lim_{i\to\infty} \delta_i = 0, \ \lim_{i\to\infty} \epsilon_i = 0,$$

a sequence $\{x_i\}_{i=0}^\infty \subset C$ *is bounded, and for each integer* $i \geq 0$,

$$l_i \in E(x_i, \delta_i, \delta_i),$$

$$x_{i+1} \in \{x_i + \beta l_i : \ \beta \in [0,1]\},$$

and

$$f(x_{i+1}) \leq f(x_i + \beta l_i) + \epsilon_i, \ \beta \in [0,1],$$

$$f(x_{i+1}) \leq f(x_i).$$

Then

$$\liminf_{i\to\infty} \Lambda(x_i) \geq 0.$$

In the following result related to algorithm $\mathcal{A}1$, we assume that the sequence of iterates has a bounded subsequence. It is proved in Sect. 3.5.

Theorem 3.12. *Assume that* $\{\delta_i\}_{i=0}^\infty \subset (0,1]$, $\{\epsilon_i\}_{i=0}^\infty \subset (0,1)$ *satisfy*

$$\lim_{i\to\infty} \delta_i = 0, \ \lim_{i\to\infty} \epsilon_i = 0,$$

a sequence $\{x_i\}_{i=0}^\infty \subset C$ *satisfies*

$$\liminf_{i\to\infty} \|x_i\| < \infty,$$

and for each integer $i \geq 0$,

$$l_i \in E(x_i, \delta_i, \delta_i),$$

$$x_{i+1} \in \{x_i + \beta l_i : \ \beta \in [0,1]\},$$

and

$$f(x_{i+1}) \leq f(x_i + \beta l_i) + \epsilon_i, \ \beta \in [0,1],$$

$$f(x_{i+1}) \leq f(x_i).$$

Then

$$\limsup_{i\to\infty} \Lambda(x_i) \geq 0.$$

The following result is related to algorithm $\mathcal{A}2$. We assume that the sequence of iterates is bounded. It is proved in Sect. 3.5.

Theorem 3.13. *Assume that* $\{\alpha_t\}_{t=0}^{\infty} \subset (0,1)$ *satisfies*

$$\lim_{i\to\infty} \alpha_i = 0, \quad \sum_{i=0}^{\infty} \alpha_i = \infty,$$

$\{\delta_i\}_{i=0}^{\infty} \subset (0,1)$ *satisfies*

$$\lim_{i\to\infty} \delta_i = 0,$$

a sequence $\{x_i\}_{i=0}^{\infty} \subset C$ *is bounded, and that for each integer* $i \geq 0$,

$$l_i \in E(x_i, \delta_i, \delta_i),$$

and if

$$f(x_i + \alpha_i l_i) < f(x_i),$$

then

$$x_{i+1} = x_i + \alpha_i l_i;$$

otherwise, $x_{i+1} = x_i$. *Then* $\limsup_{i\to\infty} \Lambda(x_i) \geq 0$.

The following result is related to algorithm $\mathcal{A}2$ too. We assume that the objective function satisfies a growth condition. It is proved in Sect. 3.5.

Theorem 3.14. *Assume that*

$$\lim_{\|x\|\to\infty,\ x\in U} f(x) = \infty,$$

$\{\alpha_i\}_{i=0}^{\infty} \subset (0,1]$ *satisfies*

$$\lim_{i\to\infty} \alpha_i = 0, \quad \sum_{i=0}^{\infty} \alpha_i = \infty,$$

and $K, \epsilon > 0$. *Then there exist* $\delta > 0$ *such that for each sequence* $\{x_i\}_{i=0}^{\infty} \subset C$ *and each sequence* $\{l_i\}_{i=0}^{\infty} \subset X$ *such that*

$$\|x_0\| \leq K$$

and for each integer $i \geq 0$,

$$l_i \in E(x_i, \delta, \delta),$$

if

$$f(x_i + \alpha_i l_i) < f(x_i),$$

then

$$x_{i+1} = x_i + \alpha_i l_i;,$$

and otherwise, $x_{i+1} = x_i$, *the inequality* $\limsup_{i\to\infty} \Lambda(x_i) \geq -\epsilon$ *holds.*

Our final result is related to algorithm $\mathcal{A}3$. It is proved in Sect. 3.6.

Theorem 3.15. *Assume that*

$$\lim_{\|x\|\to\infty,\ x\in U} f(x) = \infty$$

and $K, \epsilon > 0$. Then there exist $\beta_0 \in (0,1)$ and $\Delta > 0$ such that for each $\beta_1 \in (0, \beta_0]$, there exists a natural number N_0 such that the following assertions hold:

1. *For each sequence $\{x_i\}_{i=0}^{N_0} \subset C$, each sequence $\{l_i\}_{i=0}^{N_0-1} \subset X$, and each sequence*

$$\alpha_i \in [\beta_1, \beta_0],\ i = 0, \ldots, N_0 - 1,$$

   which satisfy

$$\|x_0\| \leq K,$$

   and for each integer $i \in \{0, \ldots, N_0 - 1\}$,

$$l_i \in E(x_i, \Delta, \Delta),$$

   and

$$x_{i+1} = x_i + \alpha_i l_i,$$

   the inequality

$$\max\{\Lambda(x_i):\ i = 0, \ldots, N_0 - 1\} \geq -\epsilon$$

   is true.

2. *For each integer $n \geq N_0$, each sequence $\{x_i\}_{i=0}^{n} \subset C$, each sequence $\{l_i\}_{i=0}^{n-1} \subset X$, and each sequence*

$$\alpha_i \in [\beta_1, \beta_0],\ i = 0, \ldots, n - 1,$$

   which satisfy

$$\|x_0\| \leq K,$$

   and for each integer $i \in \{0, \ldots, n - 1\}$,

$$l_i \in E(x_i, \Delta, \Delta),$$

   and if

$$f(x_i + \alpha_i l_i) < f(x_i),$$

   then

$$x_{i+1} = x_i + \alpha_i l_i,$$

   and otherwise, $x_{i+1} = x_i$, the inequality

$$Card(\{i \in \{0, 1, \ldots, n - 1\} : \Lambda(x_i) \leq -\epsilon\}) \leq N_0$$

   is true.

3.4 Proofs of Theorems 3.5–3.9

Proof of Theorem 3.5

Clearly, for each integer $i \geq 0$,

$$f(x_i) \leq f(x_{i+1}). \tag{3.40}$$

There exists $K > 1$ such that

$$\|x_i\| \leq K, \ i = 0, 1, \ldots \tag{3.41}$$

Assume that the theorem does not hold. Then

$$\liminf_{i \to \infty} \Lambda(x_i) < -\epsilon, \tag{3.42}$$

where $\epsilon > 0$.

Lemma 3.4 implies that there exist $\alpha, \gamma, \delta \in (0, 1)$ such that the following property holds:

(a) For each $x \in B(0, K) \cap C$ satisfying

$$\Lambda(x) \leq -\epsilon,$$

each $l \in E(x, \delta, \delta)$, and each $\beta \in (0, \alpha]$, the inequality

$$f(x) - f(x + \beta l) \geq \beta \gamma$$

holds.

By (3.42), there exists a strictly increasing sequence of natural numbers $\{i_p\}_{p=1}^{\infty}$ such that

$$\Lambda(x_{i_p}) < -\epsilon, \ p = 1, 2, \ldots \tag{3.43}$$

Property (a) and Eqs. (3.41) and (3.43) imply that for each integer $p \geq 1$,

$$f(x_{i_p+1}) - f(x_{i_p}) \leq f(x_{i_p} + \alpha l_{i_p}) - f(x_{i_p}) \leq -\alpha \gamma. \tag{3.44}$$

It follows from (3.40) and (3.44) that for each integer $n \geq 2$,

$$f(x_0) - \inf(f, C) \geq f(x_0) - f(x_{i_n+1}) = \sum_{j=0}^{i_n}(f(x_j) - f(x_{j+1}))$$

$$\geq \sum_{p=1}^{n}(f(x_{i_p}) - f(x_{i_p+1})) \geq n\alpha\gamma \to \infty$$

as $n \to \infty$. The contradiction we have reached proves Theorem 3.5.

Proof of Theorem 3.6 Assume that the theorem does not hold. Then there exists $\epsilon > 0$ such that

$$\limsup_{i\to\infty} \Lambda(x_i) < -\epsilon.$$

Hence, there exists an integer $p_0 \geq 1$ such that

$$\Lambda(x_i) < -\epsilon \text{ for every integer } i \geq p_0. \tag{3.45}$$

Since the sequence $\{x_i\}_{i=0}^{\infty}$ is bounded, there exists $K > 0$ such that

$$\|x_i\| \leq K, \ i = 0, 1, \ldots \tag{3.46}$$

Lemma 3.4 implies that there exist $\alpha, \gamma, \delta \in (0,1)$ such that the following property holds:

(a) For each $x \in B(0, K) \cap C$ satisfying

$$\Lambda(x) \leq -\epsilon,$$

each $l \in E(x, \delta, \delta)$, and each $\beta \in (0, \alpha]$, the inequality

$$f(x) - f(x + \beta l) \geq \beta\gamma$$

holds.

Since

$$\lim_{i\to\infty} \alpha_i = 0,$$

there exists an integer $p_1 \geq p_0$ such that

$$\alpha_i \leq \alpha \text{ for each integer } i \geq p_1. \tag{3.47}$$

Property (a) and Eqs. (3.45)–(3.47) imply that for each integer $i \geq p_1$,

$$f(x_i) - f(x_i + \alpha_i l_i) \geq \alpha_i \gamma, \ x_{i+1} = x_i + \alpha_i l_i,$$

$$f(x_i) - f(x_{i+1}) \geq \alpha_i \gamma. \tag{3.48}$$

It follows from (3.48) that for each integer $n \geq p_1$,

$$f(x_{p_1}) - \inf(f, C) \geq f(x_{p_1}) - f(x_n) = \sum_{i=p_1}^{n-1} (f(x_i) - f(x_{i+1}))$$

$$\geq \sum_{p_1=1}^{n-1} \alpha_i \gamma \to \infty$$

as $n \to \infty$. The contradiction we have reached proves Theorem 3.6.

Proof of Theorem 3.7 It follows from the assumptions of the theorem that there exists

$$K_0 > \sup\{f(x) : x \in U \cap B(0, K+1)\} + 1 \tag{3.49}$$

and $K_1 > 0$ such that

$$\{x \in C : f(x) \le K_0 + 1\} \subset B(0, K_1). \tag{3.50}$$

Lemma 3.4 implies that there exist $\alpha, \gamma, \delta \in (0, 1)$ such that the following property holds:

(a) For each $x \in B(0, K_1) \cap C$ satisfying

$$\Lambda(x) \le -\epsilon,$$

each $l \in E(x, \delta, \delta)$, and each $\beta \in (0, \alpha]$, the inequality

$$f(x) - f(x + \beta l) \ge \beta \gamma$$

holds.

Choose a natural number

$$N_0 > (\alpha \gamma)^{-1}(K_0 - \inf(f, C)). \tag{3.51}$$

Assume that $n \ge N_0$ is an integer, $\{x_i\}_{i=0}^n \subset C$,

$$\|x_0\| \le K, \tag{3.52}$$

and for each integer $i \in \{0, \ldots, n-1\}$,

$$l_i \in E(x_i, \delta, \delta), \ \alpha_i \in [0, 1], \tag{3.53}$$

$$f(x_i + \alpha_i l_i) \le f(x_i + \beta l), \ \beta \in [0, 1], \ x_{i+1} = x_i + \alpha_i l_i. \tag{3.54}$$

In view of (3.54), for each $i = 0, \ldots, n-1$,

$$f(x_{i+1}) \le f(x_i). \tag{3.55}$$

Equations (3.49), (3.50), (3.52), and (3.55) imply that

$$f(x_i) \le f(x_0) \le K_0, \ i = 0, 1, \ldots, n \tag{3.56}$$

and

$$\|x_i\| \le K_1, \ i = 0, \ldots, n. \tag{3.57}$$

Assume that $i \in \{0, \ldots, n-1\}$ and

$$\Lambda(x_i) \le -\epsilon. \tag{3.58}$$

Property (a) and Eqs. (3.53), (3.54), (3.57), and (3.58) imply that

$$f(x_i) - f(x_{i+1}) \ge f(x_i) - f(x_i + \alpha l_i) \ge \alpha \gamma. \tag{3.59}$$

By (3.51), (3.55), (3.56), and (3.59),

$$K_0 - \inf(f, C) \geq f(x_0) - f(x_n) = \sum_{i=0}^{n-1} (f(x_i) - f(x_{i+1}))$$

$$\geq \sum \{ f(x_i) - f(x_{i+1}) : \ i \in \{0, \dots, n-1\}, \ \Lambda(x_i) \leq -\epsilon \}$$

$$\geq \alpha\gamma \mathrm{Card}(\{ i \in \{0, \dots, n-1\}, \ \Lambda(x_i) \leq -\epsilon \})$$

and

$$\mathrm{Card}(\{ i \in \{0, \dots, n-1\}, \ \Lambda(x_i) \leq -\epsilon \})$$

$$\leq (\alpha\gamma)^{-1}(K_0 - \inf(f, C)) \leq N_0.$$

Theorem 3.7 is proved.

  Proof of Theorem 3.8

  It follows from the assumptions of the theorem that there exist

$$K_0 > \sup\{ f(x) : \ x \in U \cap B(0, K) \} \tag{3.60}$$

and $K_1 > 0$ such that

$$\{ x \in U : \ f(x) \leq K_0 + 1 \} \subset B(0, K_1). \tag{3.61}$$

Lemma 3.4 implies that there exist $\alpha, \gamma, \delta \in (0, 1)$ such that the following property holds:

(a) For each $x \in B(0, K_1) \cap C$ satisfying

$$\Lambda(x) \leq -\epsilon,$$

  each $l \in E(x, \delta, \delta)$, and each $\beta \in (0, \alpha]$, the inequality

$$f(x) - f(x + \beta l) \geq \beta\gamma$$

  holds.

  Assume that $\{\alpha_i\}_{i=0}^{\infty} \subset (0, 1]$ satisfies

$$\lim_{i \to \infty} \alpha_i = 0, \ \sum_{i=0}^{\infty} \alpha_i = \infty. \tag{3.62}$$

In view of (3.62), there exists a natural number N_0 such that

$$\alpha_i < \alpha \ \text{for every integer} \ i \geq N_0, \tag{3.63}$$

and there exists an integer $N > N_0$ such that

$$\sum_{i=N_0}^{N-1} \alpha_i > \gamma^{-1}(K_0 - \inf(f, C)). \tag{3.64}$$

Assume that $\{x_i\}_{i=0}^N \subset C$,

$$\|x_0\| \leq K, \tag{3.65}$$

and that for each integer $i \in \{0, \ldots, N-1\}$,

$$l_i \in E(x_i, \delta, \delta), \tag{3.66}$$

and

if $f(x_i + \alpha_i l_i) < f(x_i)$, then $x_{i+1} = x_i + \alpha_i l_i$; otherwise, $x_{i+1} = x_i$. $\tag{3.67}$

We show that

$$\max\{\Lambda(x_i) : i = 0, \ldots, N-1\} \geq -\epsilon.$$

Assume the contrary. Then

$$\Lambda(x_i) < -\epsilon, \; i = 0, \ldots, N-1. \tag{3.68}$$

By (3.60), (3.61), (3.65), and (3.67), for each integer $i = 0, \ldots, N-1$,

$$f(x_{i+1}) \leq f(x_i) \leq f(x_0) \leq K_0 \tag{3.69}$$

and

$$\|x_i\| \leq K_1, \; i = 0, \ldots, N. \tag{3.70}$$

Let

$$i \in \{N_0, \ldots, N-1\}. \tag{3.71}$$

Property (a) and equations (3.63), (3.66)–(3.68), (3.70) and (3.71) imply that

$$f(x_i) - f(x_i + \alpha_i l_i) \geq \alpha_i \gamma, \; x_{i+1} = x_i + \alpha_i l_i,$$

$$f(x_i) - f(x_{i+1}) \geq \alpha_i \gamma. \tag{3.72}$$

It follows from (3.69), (3.71), and (3.72) that

$$K_0 - \inf(f, C) \geq f(x_{N_0}) - f(x_N) \geq \sum_{i=N_0}^{N-1} (f(x_i) - f(x_{i+1}))$$

$$\geq \sum_{i=N_0}^{N-1} \alpha_i \gamma$$

and

$$\sum_{i=N_0}^{N-1} \alpha_i \leq \gamma^{-1}(K_0 - \inf(f, C)).$$

This contradicts (3.64). The contradiction we have reached proves Theorem 3.8.

Proof of Theorem 3.9

Fix

$$K_0 > \sup\{f(x): \ x \in U \cap B(0, K)\}. \tag{3.73}$$

There exists $K_1 > 0$ such that

$$\{x \in U: \ f(x) \leq K_0 + 1\} \subset B(0, K_1). \tag{3.74}$$

Lemma 3.4 implies that there exist $\alpha, \gamma, \delta \in (0, 1)$ such that the following property holds:

(a) For each $x \in B(0, K_1) \cap C$ satisfying

$$\Lambda(x) \leq -\epsilon,$$

each $l \in E(x, \delta, \delta)$, and each $\beta \in (0, \alpha]$, the inequality

$$f(x) - f(x + \beta l) \geq \beta \gamma$$

holds.

Set

$$\delta_0 = \alpha \gamma / 2 \tag{3.75}$$

and choose an integer

$$N_0 > 2(\alpha \gamma)^{-1}(K_0 - \inf(f, C)). \tag{3.76}$$

Assume that $\{x_i\}_{i=0}^{N_0} \subset C$,

$$\|x_0\| \leq K, \tag{3.77}$$

and for each integer $i \in \{0, \ldots, N_0 - 1\}$,

$$l_i \in E(x_i, \delta, \delta), \ \alpha_i \in [0, 1], \tag{3.78}$$

$$f(x_i + \alpha_i l_i) \leq f(x_i + \beta l) + \delta_0, \ \beta \in [0, 1], \tag{3.79}$$

and

$$x_{i+1} = x_i + \alpha_i l_i. \tag{3.80}$$

We show that

$$\max\{\Lambda(x_i): \ i = 0, \ldots, N_0 - 1\} \geq -\epsilon.$$

Assume the contrary. Then

$$\Lambda(x_i) < -\epsilon, \ i = 0, \ldots, N_0 - 1. \tag{3.81}$$

Assume that $i \in \{0, \ldots, N_0 - 1\}$ and

$$f(x_i) \leq K_0. \tag{3.82}$$

(Note that by (3.73) and (3.77), inequality (3.82) holds for $i = 0$.) Equations (3.74) and (3.82) imply that

$$\|x_i\| \leq K_1. \tag{3.83}$$

Property (a) and Eqs. (3.78), (3.81), and (3.83) imply that

$$f(x_i) - f(x_i + \alpha l_i) \geq \alpha\gamma. \tag{3.84}$$

By (3.75), (3.79), (3.80), and (3.84),

$$f(x_{i+1}) \leq f(x_i + \alpha l_i) + \delta_0$$

$$\leq f(x_i) - \alpha\gamma + \delta_0 \leq f(x_i) - \alpha\gamma/2. \tag{3.85}$$

In view of (3.84) and (3.85),

$$f(x_{i+1}) \leq K_0.$$

Therefore by induction, we showed that

$$f(x_i) \leq K_0, \; i = 0, \ldots, N_0 \tag{3.86}$$

and

$$f(x_{i+1}) \leq f(x_i) - \alpha\gamma/2, \; i = 0, \ldots, N_0 - 1. \tag{3.87}$$

It follows from (3.86) and (3.87) that

$$K_0 - \inf(f, C) \geq f(x_0) - f(x_{N_0}) = \sum_{i=0}^{N_0-1} (f(x_i) - f(x_{i+1})) \geq N_0\alpha\gamma$$

and

$$N_0 \leq 2(\alpha\gamma)^{-1}(K_0 - \inf(f, C)).$$

This contradicts (3.76). The contradiction we have reached proves that

$$\max\{\Lambda(x_i) : \; i = 0, \ldots, N_0 - 1\} \geq -\epsilon$$

and Theorem 3.9.

3.5 Proofs of Theorems 3.10–3.14

Proof of Theorem 3.10
    Fix

$$K_0 > \sup\{f(x) : \; x \in C \cap B(0, K)\}. \tag{3.88}$$

There exists $K_1 > 0$ such that

$$\{x \in C : \; f(x) \leq K_0 + 1\} \subset B(0, K_1). \tag{3.89}$$

Lemma 3.4 implies that there exist $\alpha, \gamma, \delta \in (0, 1)$ such that the following property holds:

(a) For each $x \in B(0, K_1) \cap C$ satisfying

$$\Lambda(x) \leq -\epsilon,$$

each $l \in E(x, \delta, \delta)$, and each $\beta \in (0, \alpha]$, the inequality

$$f(x) - f(x + \beta l) \geq \beta\gamma$$

holds.
Set

$$\delta_0 = \alpha\gamma/2 \tag{3.90}$$

and choose an integer

$$N_0 > 2(\alpha\gamma)^{-1}(K_0 - \inf(f, C)). \tag{3.91}$$

Assume that $\{x_i\}_{i=0}^{\infty} \subset C$,

$$\|x_0\| \leq K, \tag{3.92}$$

and for each integer $i \geq 0$,

$$l_i \in E(x_i, \delta, \delta), \tag{3.93}$$

$$x_{i+1} \in \{x_i + \beta l_i : \beta \in [0, 1]\}, \tag{3.94}$$

$$f(x_{i+1}) \leq f(x_i + \beta l_i) + \delta_0, \ \beta \in [0, 1], \tag{3.95}$$

and

$$f(x_{i+1}) \leq f(x_i). \tag{3.96}$$

By (3.88), (3.92), (3.95), and (3.96), for each integer $i \geq 0$,

$$f(x_{i+1}) \leq f(x_i) \leq f(x_0) \leq K_0 \tag{3.97}$$

and

$$\|x_i\| \leq K_1. \tag{3.98}$$

Property (a) and Eqs. (3.90), (3.93), (3.95), and (3.98) imply that the following property holds:

(b) For each integer $i \geq 0$ satisfying $\Lambda(x_i) \leq -\epsilon$, we have

$$f(x_{i+1}) \leq f(x_i + \alpha l_i) + \delta_0 \leq f(x_i) - \alpha\gamma + \delta_0 \leq f(x_i) - \alpha\gamma/2.$$

Property (b) and Eq. (3.91) and (3.97) imply that for every natural number n,

$$K_0 - \inf(f, C) \geq f(x_0) - f(x_n) = \sum_{i=0}^{n-1} (f(x_i) - f(x_{i+1}))$$

$$\geq \sum \{f(x_i) - f(x_{i+1}) : \ i \in \{0, \ldots, n-1\}, \ \Lambda(x_i) \leq -\epsilon\}$$

$$\geq 2^{-1}\alpha\gamma \mathrm{Card}(\{i \in \{0, \ldots, n-1\} : \ \Lambda(x_i) \leq -\epsilon\})$$

and

$$\mathrm{Card}(\{i \in \{0, \ldots, n-1\} : \ \Lambda(x_i) \leq -\epsilon\})$$

$$\leq 2(\alpha\gamma)^{-1}(K_0 - \inf(f, C)) < N_0.$$

This implies that

$$\mathrm{Card}(\{i \in \{0, 1, \ldots\} : \ \Lambda(x_i) \leq -\epsilon\}) \leq -N_0.$$

Theorem 3.10 is proved.

Proof of Theorem 3.11

There exists $K > 1$ such that

$$\|x_i\| \leq K, \ i = 0, 1, \ldots \tag{3.99}$$

Clearly, for each integer $i \geq 0$, we have

$$f(x_{i+1}) \leq f(x_i). \tag{3.100}$$

Assume that the theorem does not hold. Then there exist $\epsilon > 0$ and a subsequence $\{x_{i_p}\}_{p=1}^{\infty}$ such that

$$\Lambda(x_{i_p}) \leq -\epsilon, \ p = 1, 2, \ldots \tag{3.101}$$

Lemma 3.4 implies that there exist $\alpha, \gamma, \delta \in (0, 1)$ such that the following property holds:

(a) For each $x \in B(0, K) \cap C$ satisfying

$$\Lambda(x) \leq -\epsilon,$$

each $l \in E(x, \delta, \delta)$, and each $\beta \in (0, \alpha]$, the inequality

$$f(x) - f(x + \beta l) \geq \beta\gamma$$

holds.

In view of the equations

$$\lim_{i \to \infty} \delta_i = 0, \ \lim_{i \to \infty} \epsilon_i = 0,$$

there exists an integer $p_0 \geq 1$ such that for each integer $i \geq i_{p_0}$,

$$\epsilon_i \leq \alpha\gamma/4, \ \delta_i \leq \delta. \tag{3.102}$$

Let $p \geq p_0$ be an integer. Property (a), (3.99), (3.101), and (3.102) imply that

$$f(x_{i_p+1}) \leq f(x_{i_p} + \alpha l_{i_p}) + \epsilon_{i_p}$$

$$\leq f(x_{i_p}) - \alpha\gamma + \epsilon_{i_p} \leq f(x_{i_p}) - \alpha\gamma 2. \tag{3.103}$$

In view of (3.100) and (3.103), for every integer $n \geq p_0 + 1$,

$$f(x_0) - \inf(f, C) \geq f(x_0) - f(x_{i_n})$$

$$= \sum_{j=0}^{i_n-1} (f(x_i) - f(x_{i+1}))$$

$$\geq \sum_{i=p_0}^{n-1} (f(x_{i_p}) - f(x_{i_p+1})) \geq 2^{-1}\alpha\gamma(n - 1 - p_0) \to \infty$$

as $n \to \infty$. The contradiction we have reached proves Theorem 3.11.

Proof of Theorem 3.12

Assume that the theorem does not hold. Then there exists $\epsilon > 0$ such that

$$\limsup_{i \to \infty} \Lambda(x_i)) < -\epsilon.$$

We may assume without loss of generality that

$$\Lambda(x_i) < -\epsilon, \ i = 0, 1, \ldots \tag{3.104}$$

There exist $K > 0$ and a subsequence $\{x_{i_p}\}_{p=1}^{\infty}$ such that

$$\|x_{i_p}\| \leq K, \ p = 1, 2, \ldots \tag{3.105}$$

Lemma 3.4 implies that there exist $\alpha, \gamma, \delta \in (0, 1)$ such that the following property holds:

(a) For each $x \in B(0, K) \cap C$ satisfying

$$\Lambda(x) \leq -\epsilon,$$

each $l \in E(x, \delta, \delta)$, and each $\beta \in (0, \alpha]$, the inequality

$$f(x) - f(x + \beta l) \geq \beta\gamma$$

holds.

There exists an integer $i_0 \geq 1$ such that for each integer $i \geq i_0$,

$$\epsilon_i < \alpha\gamma/2, \ \delta_i \leq \delta. \tag{3.106}$$

Let $p \geq 1$ be an integer satisfying $i_p \geq i_0$. Property (a) and (3.104)–(3.106) imply that

$$f(x_{i_p}) - f(x_{i_p} + \alpha l_{i_p}) \geq \alpha \gamma. \tag{3.107}$$

By (3.106) and (3.107),

$$f(x_{i_p+1}) \leq f(x_{i_p} + \alpha l_{i_p}) + \epsilon_{i_p}$$

$$\leq f(x_{i_p}) - \alpha \gamma + \epsilon_{i_p} \leq f(x_{i_p}) - \alpha \gamma 2. \tag{3.108}$$

Let $p \geq 1$ be an integer satisfying $i_p \geq i_0$. By (3.108), for each integer $n > p$,

$$f(x_0) - \inf(f, C) \geq f(x_0) - f(x_{i_n+1})$$

$$= \sum_{j=0}^{i_n} (f(x_j) - f(x_{j+1}))$$

$$\geq \sum_{j=i_p}^{i_n} (f(x_{i_p}) - f(x_{i_p+1})) \geq 2^{-1} \alpha \gamma (n - p) \to \infty$$

as $n \to \infty$. The contradiction we have reached proves Theorem 3.12.

Proof of Theorem 3.13

There exists $K > 0$ such that

$$\|x_i\| \leq K, \; i = 0, 1, \ldots \tag{3.109}$$

Assume that the theorem does not hold. Then there exists $\epsilon > 0$ such that

$$\limsup_{i \to \infty} \Lambda(x_i)) < -\epsilon.$$

We may assume without loss of generality that

$$\Lambda(x_p) < -\epsilon, \; p = 0, 1, \ldots \tag{3.110}$$

Lemma 3.4 implies that there exist $\alpha, \gamma, \delta \in (0, 1)$ such that the following property holds:

(a) For each $x \in B(0, K) \cap C$ satisfying

$$\Lambda(x) \leq -\epsilon,$$

each $l \in E(x, \delta, \delta)$, and each $\beta \in (0, \alpha]$, the inequality

$$f(x) - f(x + \beta l) \geq \beta \gamma$$

holds.

Clearly, for each integer $i \geq 0$,

$$f(x_{i+1}) \leq f(x_i). \tag{3.111}$$

There exists an integer $p_0 \geq 1$ such that for each integer $i \geq p_0$,

$$\alpha_i < \alpha, \ \delta_i < \delta. \tag{3.112}$$

Property (a) and (3.109), (3.110), and (3.112) imply that for each integer $i \geq p_0$,

$$f(x_i) - f(x_i + \alpha_i l_i) \geq \alpha_i \gamma. \tag{3.113}$$

By (3.113) for each integer $n > p_0$,

$$f(x_{p_0}) - \inf(f, C) \geq f(x_{p_0}) - f(x_n)$$

$$= \sum_{i=p_0}^{n-1} (f(x_i) - f(x_{i+1})) \geq \gamma \sum_{i=p_0}^{n-1} \alpha_i \to \infty$$

as $n \to \infty$. The contradiction we have reached proves Theorem 3.13.

Proof of Theorem 3.14

Fix

$$K_0 > \sup\{f(x) : \ x \in U \cap B(0, K)\} + 1. \tag{3.114}$$

There exists $K_1 > 0$ such that

$$\{x \in U : \ f(x) \leq K_0 + 1\} \subset B(0, K_1). \tag{3.115}$$

Lemma 3.4 implies that there exist $\alpha, \gamma, \delta \in (0, 1)$ such that the following property holds:

(a) For each $x \in B(0, K_1) \cap C$ satisfying

$$\Lambda(x) \leq -\epsilon,$$

each $l \in E(x, \delta, \delta)$, and each $\beta \in (0, \alpha]$, the inequality

$$f(x) - f(x + \beta l) \geq \beta \gamma$$

holds.

Assume that $\{x_i\}_{i=0}^{\infty} \subset C$,

$$\|x_0\| \leq K, \tag{3.116}$$

and that for each integer $i \geq 0$,

$$l_i \in E(x_i, \delta, \delta); \tag{3.117}$$

if $f(x_i + \alpha_i l_i) < f(x_i)$, then $x_{i+1} = x_i + \alpha_i l_i$; otherwise, $x_{i+1} = x_i$. (3.118)

We show that
$$\limsup_{i \to \infty} \Lambda(x_i) \geq -\epsilon.$$

Assume the contrary. Then
$$\limsup_{i \to \infty} \Lambda(x_i) < -\epsilon.$$

There exists an integer $i_0 \geq 1$ such that
$$\Lambda(x_i) < -\epsilon \text{ for every integer } i \geq i_0. \tag{3.119}$$

By (3.114), (3.116), and (3.118), for each integer $i \geq 0$,
$$f(x_{i+1}) \leq f(x_i) \leq f(x_0) \leq K_0. \tag{3.120}$$

In view of (3.115) and (3.120),
$$\|x_i\| \leq K_1, \ i = 0, 1, \ldots \tag{3.121}$$

There exists an integer $i_1 > i_0$ such that for each integer $i \geq i_1$,
$$\alpha_i < \alpha. \tag{3.122}$$

Let $i \geq i_1$ be an integer. Property (a) and Eqs. (3.117), (3.119), and (3.121) imply that
$$f(x_i) - f(x_i + \alpha_i l_i) \geq \alpha_i \gamma.$$

Together with (3.118), this implies that
$$x_{i+1} = x_i + \alpha_i x_i$$

and
$$f(x_i) - f(x_{i+1}) \geq \alpha_i \gamma. \tag{3.123}$$

By (3.123), for each integer $n > i_i$,
$$f(x_{i_1}) - \inf(f, C) \geq f(x_{i_1}) - f(x_n) = \sum_{i=i_1}^{n-1} (f(x_i) - f(x_{i+1}))$$

$$\geq \gamma \sum_{i=i_1}^{n-1} \alpha_i \to \infty$$

as $n \to \infty$. The contradiction we have reached proves that
$$\limsup_{i \to \infty} \Lambda(x_i) \geq -\epsilon$$

and Theorem 3.14.

3.6 Proof of Theorem 3.15

Fix
$$K_0 > \sup\{f(x) : \ x \in U \cap B(0, K)\}. \tag{3.124}$$

There exists $K_1 > K$ such that

$$\{x \in U : \ f(x) \le K_0 + 1\} \subset B(0, K_1). \tag{3.125}$$

Lemma 3.4 implies that there exist $\beta_0, \gamma, \Delta \in (0, 1)$ such that the following property holds:

(a) For each $x \in B(0, K_1) \cap C$ satisfying

$$\Lambda(x) \le -\epsilon,$$

each $l \in E(x, \Delta, \Delta)$, and each $\beta \in (0, \beta_0]$, the inequality

$$f(x) - f(x + \beta l) \ge \beta \gamma$$

holds.

Assume that
$$\beta_1 \in (0, \beta_0]. \tag{3.126}$$

Choose a natural number

$$N_0 > (\beta_1 \gamma)^{-1}(K_0 - \inf(f, C)). \tag{3.127}$$

Let us prove assertion 1. Assume that $\{x_i\}_{i=0}^{N_0} \subset C$,

$$\|x_0\| \le K, \tag{3.128}$$

$\{l_i\}_{i=0}^{N_0-1} \subset X$,
$$\alpha_i \in [\beta_1, \beta_0], \ i = 0, \ldots, N_0 - 1, \tag{3.129}$$

and for each integer $i \in \{0, \ldots, N_0 - 1\}$,

$$l_i \in E(x_i, \Delta, \Delta) \tag{3.130}$$

and

$$x_{i+1} = x_i + \alpha_i l_i. \tag{3.131}$$

In order to prove assertion 1, it is sufficient to show that

$$\max\{\Lambda(x_i) : \ i = 0, \ldots, N_0 - 1\} \ge -\epsilon.$$

Assume the contrary. Then

$$\Lambda(x_i) < -\epsilon, \ i = 0, \ldots, N_0 - 1. \tag{3.132}$$

Assume that $i \in \{0, \ldots, N_0 - 1\}$ and

$$f(x_i) \leq K_0. \tag{3.133}$$

(Note that in view of (3.124) and (3.128), our assumption holds for $i = 0$.)
By (3.125) and (3.133),
$$\|x_i\| \leq K_1. \tag{3.134}$$

Property (a) and Eqs. (3.129)–(3.132) and (3.134) imply that

$$f(x_{i+1}) = f(x_i + \alpha_i l_i) \leq f(x_i) - \alpha_i \gamma \leq f(x_i) - \beta_1 \gamma. \tag{3.135}$$

It follows from (3.133) and (3.135) that

$$f(x_{i+1}) \leq f(x_i) \leq K_0.$$

Therefore by induction, we proved that for all integers $i = 0, \ldots, N_0$,

$$f(x_i) \leq K_0 \tag{3.136}$$

and that for all integers $i = 0, \ldots, N_0 - 1$,

$$f(x_{i+1}) \leq f(x_i) - \beta_1 \gamma. \tag{3.137}$$

By (3.136) and (3.137),

$$K_0 - \inf(f, C) \geq f(x_0) - f(x_{N_0}) = \sum_{i=0}^{N_0 - 1} (f(x_i) - f(x_{i+1})) \geq N_0 \beta_1 \gamma$$

and

$$N_0 \leq (\beta_1 \gamma)^{-1}(K_0 - \inf(f, C)).$$

This contradicts (3.127). The contradiction we have reached proves assertion 1.

Let us prove assertion 2. Assume that $n \geq N_0$ is an integer, $\{x_i\}_{i=0}^{n} \subset C$, each sequence $\{l_i\}_{i=0}^{n-1} \subset X$,

$$\alpha_i \in [\beta_1, \beta_0], \; i = 0, \ldots, n - 1, \tag{3.138}$$

$$\|x_0\| \leq K, \tag{3.139}$$

and for each integer $i \in \{0, \ldots, n - 1\}$,

$$l_i \in E(x_i, \Delta, \Delta) \tag{3.140}$$

and

if $f(x_i + \alpha_i l_i) < f(x_i)$, then $x_{i+1} = x_i + \alpha_i l_i$; otherwise, $x_{i+1} = x_i$. (3.141)

In order to prove assertion 2, it is sufficient to show that

$$\text{Card}(\{i \in \{0, 1, \ldots, n-1\} : \Lambda(x_i) < -\epsilon\}) \leq N_0.$$

By (3.124) and (3.139),
$$f(x_0) \leq K_0.$$

In view of (3.141), for each integer $i \in \{0, \ldots, n-1\}$,

$$f(x_{i+1}) \leq f(x_i) \leq K_0.$$

Together with (3.125), this implies that for all integers $i \in \{0, \ldots, n\}$,

$$f(x_i) \leq K_0, \quad \|x_i\| \leq K_1. \tag{3.142}$$

Assume that $i \in \{0, \ldots, n-1\}$ and

$$\Lambda(x_i) < -\epsilon. \tag{3.143}$$

Property (a), (3.138), (3.140), (3.142), and (3.143) imply that

$$f(x_i + \alpha_i l_i) \leq f(x_i) - \alpha_i \gamma \leq f(x_i) - \beta_1 \gamma. \tag{3.144}$$

It follows from (3.141), (3.142), and (3.144) that

$$K_0 - \inf(f, C) \geq f(x_0) - f(x_n) = \sum_{i=0}^{n-1} (f(x_i) - f(x_{i+1}))$$

$$\geq \sum \{f(x_i) - f(x_{i+1}) : i \in \{0, 1, \ldots, n-1\}, \Lambda(x_i) < -\epsilon\}$$

$$\geq -\beta_1 \gamma \text{Card}(\{i \in \{0, 1, \ldots, n-1\} : \Lambda(x_i) < -\epsilon\}).$$

Together with (3.127), this implies that

$$\text{Card}(\{i \in \{0, 1, \ldots, n-1\} : \Lambda(x_i) < -\epsilon\})$$

$$\leq (\beta_1 \gamma)^{-1}(K_0 - \inf(f)) \leq N_0.$$

Assertion 2 is proved. This completes the proof of Theorem 3.15.

3.7 A Convergence Result with Estimations

Assume that $(X, \|\cdot\|)$ is a Banach space, $U \subset X$ is an open convex set,

$$C \subset U$$

is a nonempty closed and convex set, and a function $f : U \to R^1$ is Frechet differentiable at any point $x \in U$ and is Lipschitz on all bounded sets in U.

Recall that for each $x \in C$,

$$\Lambda(x) = \inf\{\langle f'(x), h \rangle : h \in (C - x) \cap B(0,1)\}.$$

In this section, we continue to consider the minimization problem

$$f(x) \to \min, \ x \in C.$$

In the previous sections, we showed that for a given $\epsilon > 0$, there exist a sufficiently small error $\delta, \Delta > 0$ such that after a certain number of iterations, we obtain a point $x_s \in K$ satisfying $\Lambda(x_s) > -\epsilon$. Of course, it is interesting to obtain an explicit estimation for δ, Δ. It is done in the two theorems below under some additional assumptions on the objective function f. Namely, we assume that its Frechet derivative is Holder continuous on bounded sets and that the corresponding constants are known. In this case, we obtain an explicit dependence δ, Δ and a number of iterations on ϵ. This dependence allows us to easily solve an inverse problem: if we know δ, Δ what ϵ can be obtained?

Let $K \geq 1$, $c_* > 1$ and $\lambda \in (0, 1]$ be fixed.

Recall that for each $x \in C$ and each pair of numbers $\delta, \Delta \in (0, 1]$, we denote by $E(x, \delta, \Delta)$ the set of all $l \in C - x$ for which there exist $g \in X^*$ such that

$$\|g - f'(x)\|_* \leq \delta$$

and

$$\xi \in B(0, c_*) \cap B(l, \delta)$$

such that

$$\langle g, \xi \rangle \leq \lambda \inf\{\langle g, \eta \rangle : \eta \in B(0, c_*) \cap (C - x)\} + \Delta.$$

Assume that
$$\inf(f, C) = \inf(f, C \cap B(0, K)), \tag{3.145}$$

$$K_0 > \sup\{|f(x)| : x \in U \cap B(0, K)\}, \tag{3.146}$$

$$K_1 \geq \sup\{\|x\| : x \in U, \ |f(x)| \leq K_0 + 1\}, \tag{3.147}$$

$L_0 \geq 1$, $\rho \in (0, 1]$ and that

$$|f(z_1) - f(z_2)| \leq L_0\|z_1 - z_2\|, \ z_1, z_2 \in B(0, K_1 + 1) \cap U, \tag{3.148}$$

$$\|f'(z_1) - f'(z_2)\|_* \leq L_0\|z_1 - z_2\|^\rho, \ z_1, z_2 \in B(0, K_1 + 1) \cap U. \tag{3.149}$$

Let $\epsilon \in (0, 1]$,

$$\Delta = 8^{-1}\epsilon\lambda, \tag{3.150}$$

$$\delta = 8^{-1}\epsilon\lambda(2c_* + L_0)^{-1}, \tag{3.151}$$

$$\alpha = \min\{((32c_*^2 L_0)^{-1}\epsilon\lambda)^{1/\rho},\ (2c_*)^{-1}\}. \tag{3.152}$$

In this chapter, we prove the following two results.

Theorem 3.16. *Assume that*

$$\bar{\alpha} \in (0, \alpha],$$

an integer N_0 satisfies

$$N_0 > 4K_0(\lambda\epsilon\bar{\alpha})^{-1}, \tag{3.153}$$

$\{x_i\}_{i=0}^{N_0} \subset C$, $\{l_i\}_{i=0}^{N_0-1} \subset X$,

$$\alpha_i \in [\bar{\alpha}, \alpha],\ i = 0, \ldots, N_0 - 1, \tag{3.154}$$

$$\|x_0\| \le K, \tag{3.155}$$

and that for each integer $i \in \{0, \ldots, N_0 - 1\}$,

$$l_i \in E(x_i, \delta, \Delta) \tag{3.156}$$

and

$$x_{i+1} = x_i + \alpha_i l_i. \tag{3.157}$$

Then the inequality

$$\max\{\Lambda(x_i):\ i = 0, \ldots, N_0 - 1\} \ge -\epsilon$$

holds, and there exists $j \in \{0, \ldots, N_0 - 1\}$ such that

$$f(x_j) - f(x_{j+1}) < 2^{-1}\lambda\bar{\alpha}\epsilon. \tag{3.158}$$

Moreover, if $j \in \{0, \ldots, N_0 - 1\}$ satisfies (3.158), and for each $i \in \{0, \ldots, j\} \setminus \{j\}$,

$$f(x_i) - f(x_{i+1}) \ge 2^{-1}\lambda\bar{\alpha}\epsilon,$$

then $\Lambda(x_j) \ge -\epsilon$.

Theorem 3.17. *Assume that*

$$\bar{\alpha} \in (0, \alpha],$$

an integer N_0 satisfies

$$N_0 > 4K_0(\lambda\epsilon\bar{\alpha})^{-1},$$

$n \ge N_0$ is an integer, $\{x_i\}_{i=0}^{n} \subset C$, $\{l_i\}_{i=0}^{n-1} \subset X$,

$$\alpha_i \in [\bar{\alpha}, \alpha], \ i = 0, \ldots, n-1, \tag{3.159}$$

$$\|x_0\| \leq K, \tag{3.160}$$

and that for each integer $i \in \{0, \ldots, n-1\}$,

$$l_i \in E(x_i, \delta, \Delta) \tag{3.161}$$

and

if $f(x_i + \alpha_i l_i) < f(x_i)$, then $x_{i+1} = x_i + \alpha_i l_i$; otherwise, $x_{i+1} = x_i$. (3.162)

Then

$$Card(\{i \in \{0, 1, \ldots, n-1\} : \Lambda(x_i) \leq -\epsilon\}) \leq N_0.$$

Theorems 3.16 and 3.17 are proved in Sect. 3.9. Their proofs are based on a lemma that is proved in the next section.

3.8 An Auxiliary Result

Lemma 3.18. *Let*

$$x \in B(0, K_1) \cap C \tag{3.163}$$

satisfy

$$\Lambda(x) \leq -\epsilon, \tag{3.164}$$

$$l \in E(x, \delta, \Delta), \tag{3.165}$$

$$\beta \in (0, \alpha]. \tag{3.166}$$

Then

$$f(x) - f(x + \beta l) \geq 2^{-1} \lambda \epsilon \beta$$

holds.

Proof. It follows from (3.165) that there exist

$$\xi \in B(0, c_*) \cap B(l, \delta) \tag{3.166}$$

and $g \in X^*$ such that

$$\|g - f'(x)\|_* \leq \delta \tag{3.167}$$

and

$$\langle g, \xi \rangle \leq \lambda \inf\{\langle g, \eta \rangle : \eta \in B(0, c_*) \cap (C - x)\} + \Delta. \tag{3.168}$$

By the mean value theorem, there exists $\theta \in [0, \beta]$ such that

$$f(x + \beta l) - f(x) = \langle f'(x + \theta l), \beta l \rangle. \tag{3.169}$$

Clearly,

$$\langle f'(x + \theta l), l \rangle \le \langle f'(x), l \rangle + \|f'(x + \theta l) - f'(x)\|_* \|l\|. \tag{3.170}$$

Equations (3.160), (3.163), and (3.166) imply that

$$\|l\| \le \|\xi\| + 1 \le c_* + 1 \le 2c_*,$$

$$\|\theta l\| \le 2\alpha c_*, \quad \|x + \theta l\| \le K_1 + 1. \tag{3.171}$$

By (3.149), (3.163), and (3.171),

$$\|f'(x + \theta l) - f'(x)\|_* \le L_0 \|\theta l\|^\rho \le L_0(2\alpha c_*)^\rho \le 2c_* L_0 \alpha^\rho. \tag{3.172}$$

It follows from (3.170)–(3.172) that

$$\langle f'(x + \theta l), l \rangle \le \langle f'(x), l \rangle + 4c_*^2 L_0 \alpha^\rho. \tag{3.173}$$

Lemma 3.1 and Eqs. (3.148), (3.163), (3.166)–(3.168) imply that

$$\langle f'(x), l \rangle \le \langle f'(x), \xi \rangle + \|f'(x)\|_* \|\xi - l\|$$

$$\le \langle f'(x), \xi \rangle + L_0 \delta$$

$$\le \langle g, \xi \rangle + \|f'(x) - g\|_* \|\xi\| + L_0 \delta$$

$$\le \langle g, \xi \rangle + \delta c_* + L_0 \delta$$

$$\le \delta(c_* + L_0) + \lambda \inf\{\langle g, \eta \rangle : \eta \in B(0, c_*) \cap (C - x)\} + \Delta$$

$$\le \delta(c_* + L_0) + \Delta + \lambda(\inf\{\langle f'(x), \eta \rangle : \eta \in B(0, c_*) \cap (C - x)\} + \delta c_*)$$

$$\le \delta(2c_* + L_0) + \Delta + \lambda \inf\{\langle f'(x), \eta \rangle : \eta \in B(0, c_*) \cap (C - x)\}$$

$$\le \Delta + \delta(2c_* + L_0) + \lambda \Lambda(x) \le \Delta + \delta(2c_* + L_0) - \lambda \epsilon. \tag{3.174}$$

Equations (3.150)–(3.152), (3.169), (3.173), and (3.174) imply that

$$f(x + \beta l) - f(x) \le \beta \langle f'(x), l \rangle + 4\beta c_*^2 L_0 \alpha^\rho$$

$$\le \beta(-\lambda \epsilon + \Delta + \delta(2c_* + L_0) + 4c_*^2 L_0 \alpha^\rho) \le -2^{-1} \beta \lambda \epsilon.$$

Lemma 3.18 is proved.

3.9 Proofs of Theorems 3.16 and 3.17

Proof of Theorem 3.16

Assume that the theorem does not hold. Then

$$A(x_i) < -\epsilon, \ i = 0, \ldots, N_0 - 1. \tag{3.175}$$

Assume that $i \in \{0, \ldots, N_0 - 1\}$ and

$$f(x_i) \leq K_0. \tag{3.176}$$

(Note that in view of (3.146) and (3.155), inequality (3.176) holds for $i = 0$.)
By (3.147) and (3.176),

$$\|x_i\| \leq K_1. \tag{3.177}$$

Lemma 3.18 and Eqs. (3.154), (3.156), (3.157), and (3.175)–(3.177) imply that

$$f(x_i) - f(x_{i+1}) = f(x_i) - f(x_i + \alpha_i l_i) \geq 2^{-1}\lambda\epsilon\bar{\alpha}. \tag{3.178}$$

It follows from (3.176) and (3.178) that

$$f(x_{i+1}) \leq f(x_i) \leq K_0.$$

Thus by induction we showed that

$$f(x_i) \leq K_0, \ i = 0, \ldots, N_0 \tag{3.179}$$

and for all integers $i = 0, \ldots, N_0 - 1$,

$$f(x_i) - f(x_{i+1}) \geq 2^{-1}\lambda\epsilon\bar{\alpha}. \tag{3.180}$$

By (3.145), (3.146), (3.179), and (3.180),

$$2K_0 \geq f(x_0) - \inf(f, K) \geq f(x_0) - f(x_{N_0})$$

$$= \sum_{i=0}^{N_0-1} (f(x_i) - f(x_{i+1})) \geq 2^{-1}\lambda\epsilon\bar{\alpha}N_0$$

and

$$N_0 \leq 4K_0(\lambda\epsilon\bar{\alpha})^{-1}.$$

This contradicts (3.153). The contradiction we have reached proves

$$\max\{A(x_i): \ i = 0, \ldots, N_0 - 1\} \geq -\epsilon.$$

Now assume that for each $i \in \{0, \ldots, N_0 - 1\}$,

$$f(x_i) - f(x_{i+1}) \geq 2^{-1}\lambda\epsilon\bar{\alpha}.$$

Together with (3.145), (3.146), and (3.155), this implies that

$$2K_0 \geq f(x_0) - \inf(f, C) \geq f(x_0) - f(x_{N_0})$$

$$= \sum_{i=0}^{N_0-1} (f(x_i) - f(x_{i+1})) \geq 2^{-1}\lambda\epsilon\bar{\alpha}N_0$$

and

$$N_0 \leq 4K_0(\lambda\epsilon\bar{\alpha})^{-1}.$$

This contradicts (3.153). The contradiction we have reached proves that there exists

$$j \in \{0, \ldots, N_0 - 1\}$$

such that

$$f(x_j) - f(x_{j+1}) < 2^{-1}\lambda\epsilon\bar{\alpha}, \tag{3.181}$$

and for each $i \in \{0, \ldots, j\}\{j\}$,

$$f(x_i) - f(x_{i+1}) \geq 2^{-1}\lambda\epsilon\bar{\alpha}.$$

By the equation above and (3.145), (3.146), and (3.155),

$$f(x_j) \leq f(x_0) \leq K_0.$$

Combined with (3.147), this implies that

$$\|x_j\| \leq K_1. \tag{3.182}$$

Lemma 3.18 and Eqs. (3.153), (3.156), (3.157), (3.181), and (3.182) imply that

$$\Lambda(x_j) > -\epsilon.$$

Theorem 3.16 is proved.

Proof of Theorem 3.17

By (3.146), (3.147), (3.160), and (3.162), for each integer $i \in \{0, \ldots, n-1\}$,

$$f(x_{i+1}) \leq f(x_i) \leq f(x_0) \leq K_0 \tag{3.183}$$

and

$$\|x_i\| \leq K_1, \ i = 0, 1, \ldots, n. \tag{3.184}$$

Assume that $i \in \{0, \ldots, n-1\}$ and

$$\Lambda(x_i) \leq -\epsilon. \tag{3.185}$$

Lemma 3.18 and Eqs. (3.159), (3.161), (3.162), (3.184), and (3.185) imply that

$$f(x_i) - f(x_{i+1}) = f(x_i) - f(x_i + \alpha_i l_i)$$

$$\geq 2^{-1}\lambda\epsilon\alpha_i \geq 2^{-1}\lambda\epsilon\bar{\alpha}. \tag{3.186}$$

It follows from (3.145)–(3.147), (3.166), (3.183), (3.185), and (3.186) that

$$2K_0 \geq f(x_0) - \inf(f, K) \geq f(x_0) - f(x_n)$$

$$= \sum_{i=0}^{n-1}(f(x_i) - f(x_{i+1}))$$

$$\geq \sum\{f(x_i) - f(x_{i+1}) : i \in \{0, \ldots, n-1\}, \Lambda(x_i) \leq -\epsilon\}$$

$$\geq 2^{-1}\lambda\epsilon\bar{\alpha}\mathrm{Card}(\{i \in \{0, 1, \ldots, n-1\} : \Lambda(x_i) \leq -\epsilon\})$$

and

$$\mathrm{Card}(\{i \in \{0, 1, \ldots, n-1\} : \Lambda(x_i) \leq -\epsilon\} \leq 4K_0(\lambda\epsilon\bar{\alpha})^{-1}) \leq N_0.$$

Theorem 3.17 is proved.

4

Continuous Algorithms

In this chapter, we study continuous analogs of algorithms for minimization of Frechet differentiable functions studied in Chaps. 2 and 3, under the presence of computational errors. We show that our algorithms generate a good approximate solution, if computational errors are bounded from above by a small positive constant.

4.1 Banach Space Valued Functions

Let $(X, \|\cdot\|)$ be a Banach space and $a < b$ be real numbers. For any set $E \subset R^1$, define

$$\chi_E(t) = 1 \text{ for all } t \in E \text{ and } \chi_E(t) = 0 \text{ for all } t \in R^1 \setminus E.$$

If a set $E \subset R^1$ is Lebesgue measurable, then its Lebesgue measure is denoted by $|E|$ or mes(E).

A function $f : [a, b] \to X$ is called a simple function if there exists a finite collection of Lebesgue measurable sets $E_i \subset [a, b]$, $i \in I$, mutually disjoint, and $x_i \in X$, $i \in I$ such that

$$f(t) = \sum_{i \in I} \chi_{E_i}(t) x_i, \ t \in [a, b].$$

A function $f : [a, b] \to X$ is strongly measurable if there exists a sequence of simple functions $\phi_k : [a, b] \to X$, $k = 1, 2, \ldots$ such that

$$\lim_{k \to \infty} \|\phi_k(t) - f(t)\| = 0, \ t \in [a, b] \text{ almost everywhere (a. e.).} \tag{4.1}$$

© The Author(s), under exclusive license to Springer Nature Switzerland AG 2022
A. J. Zaslavski, *Optimization in Banach Spaces*, SpringerBriefs in Optimization, https://doi.org/10.1007/978-3-031-12644-4_4

For every simple function $f(\cdot) = \sum_{i \in I} \chi_{E_i}(\cdot) x_i$, where the set I is finite, define its Bochner integral by

$$\int_a^b f(t)dt = \sum_{i \in I} |E_i| x_i.$$

Let $f : [a, b] \to X$ be a strongly measurable function. We say that f is Bochner integrable if there exists a sequence of simple functions $\phi_k : [a, b] \to X$, $k = 1, 2, \ldots$ such that (4.1) holds and the sequence $\{\int_a^b \phi_k(t)dt\}_{k=1}^\infty$ strongly converges in X. In this case, we define the Bochner integral of the function f by

$$\int_a^b f(t)dt = \lim_{k \to \infty} \int_a^b \phi_k(t)dt.$$

It is known that the integral defined above is independent of the choice of the sequence $\{\phi_k\}_{k=1}^\infty$ [27]. Similar to the Lebesgue integral, for any measurable set $E \subset [a, b]$, the Bochner integral of f over E is defined by

$$\int_E f(t)dt = \int_a^b \chi_E(t)f(t)dt.$$

If $x : [a, b] \to X$ is a Bochner integrable function, then for almost every (a. e.) $t \in [a, b]$,

$$\lim_{\Delta t \to \infty} (\Delta t)^{-1} \int_t^{t+\Delta} \|x(\tau) - x(t)\| d\tau = 0,$$

and the function

$$y(t) = \int_a^t x(s)ds, \ t \in [a, b]$$

is continuous and a.e. differentiable on $[a, b]$.

The following result is true (see Proposition 3.4, Chapter 2 of [27]).

Proposition 4.1. *Let $f : [a, b] \to X$ be a strongly measurable function. Then f is Bochner integrable if and only if the function $\|f(\cdot)\|$ is Lebesgue integrable. Moreover, in this case,*

$$\left\| \int_a^b f(t)dt \right\| \leq \int_a^b \|f(t)\| dt.$$

The Bochner integral possesses almost the same properties as the Lebesgue integral. If $f : [a, b] \to X$ is strongly measurable and $\|f(\cdot)\| \in L^p(a, b)$, for some $p \in [1, \infty)$, then we say that $f(\cdot)$ is L^p Bochner integrable. For every $p \geq 1$, the set of all L^p Bochner integrable functions is denoted by $L^p(a, b; X)$, and for every $f \in L^p(a, b; X)$,

$$\|f\|_{L^p(a,b;X)} = \left(\int_a^b \|f(t)\|^p dt \right)^{1/p}.$$

Clearly, the set of all Bochner integrable functions on $[a, b]$ is $L^1(a, b; X)$.

Let $a < b$ be real numbers. A function $x : [a, b] \to X$ is absolutely continuous (a. c.) on $[a, b]$ if for each $\epsilon > 0$, there exists $\delta > 0$ such that for each pair of sequences $\{t_n\}_{n=1}^q$, $\{s_n\}_{n=1}^q \subset [a, b]$ satisfying

$$t_n < s_n, \ n = 1, \ldots, q, \ \sum_{n=1}^q (s_n - t_n) \le \delta,$$

$$(t_n, s_n) \cap (t_m, s_m) = \emptyset \text{ for all } m, n \in \{1, \ldots, q\} \text{ such that } m \ne n,$$

we have

$$\sum_{n=1}^q \|x(t_n) - x(s_m)\| \le \epsilon.$$

The following result is true (see Theorem 1.124 of [4]).

Proposition 4.2. *Let X be a reflexive Banach space. Then every a. c. function $x : [a, b] \to X$ is a. e. differentiable on $[a, b]$, and*

$$x(t) = x(a) + \int_a^t (dx/dt)(s)ds, \ t \in [a, b],$$

where $dx/dt \in L^1(a, b; X)$ is the strong derivative of x.

Let $-\infty < \tau_1 < \tau_2 < \infty$. Denote by $W^{1,1}(\tau_1, \tau_2)$ the set of all functions $x : [\tau_1, \tau_2] \to X$ for which there exists a Bochner integrable function $u : [\tau_1, \tau_2] \to X$ such that

$$x(t) = x(\tau_1) + \int_{\tau_1}^t u(s)ds, \ t \in (\tau_1, \tau_2].$$

It is known that if $x \in W^{1,1}(\tau_1, \tau_2)$, then this equation defines a unique Bochner integrable function that is called the derivative of x and is denoted by x' [4, 8].

Lemma 4.3. *Assume that $U \subset X$ is an open convex set, $-\infty < \tau_1 < \tau_2 < \infty$, a function $x \in W^{1,1}(\tau_1, \tau_2)$ is Lipschitz, $x(t) \in U$, $t \in [\tau_1, \tau_2]$, and a function $f : U \to R^1$ is Lipschitz on all bounded subsets of U and is Frechet differentiable at all points $x(t)$ for any $t \in [\tau_1, \tau_2]$. Then the function $f \circ x$ is Lipschitz on $[\tau_1, \tau_2]$, and for a. e. $t \in [\tau_1, \tau_2]$,*

$$(f \circ x)'(t) = \lim_{\Delta t \to 0} (\Delta t)^{-1}(f(x(t) + \Delta t x'(t)) - f(x(t))) = \langle f'(x(t)), x'(t) \rangle.$$

Proof. Since the function f is Lipschitz on bounded sets, the function $f \circ x$ is Lipschitz on $[\tau_1, \tau_2]$. For a. e. $t \in [\tau_1, \tau_2]$, there exist

$$(f \cdot x)'(t) = \lim_{\Delta t \to 0} \Delta t^{-1}(f(x(t + \Delta t)) - f(x(t)))$$

and

$$x'(t) = \lim_{\Delta t \to 0} \Delta t^{-1}(x(t + \Delta t) - x(t)).$$

Let $t \in [\tau_1, \tau_2]$ and $(f \cdot x)'(t)$ and $x'(t)$ exist. Since f is Lipschitz on bounded sets, we have

$$(\Delta t)^{-1}(f(x(t + \Delta t)) - f(x(t) + \Delta t x'(t))) \to 0 \text{ as } \Delta t \to 0.$$

This implies that

$$(f \circ x)'(t) = \lim_{\Delta t \to 0}((\Delta t)^{-1}(f(x(t) + \Delta t x'(t)) - f(x(t))) = \langle f'(x(t)), x'(t)\rangle.$$

Lemma 4.3 is proved.

4.2 Convex Problems

Assume that $(X, \|\cdot\|)$ is a Banach space, $U \subset X$ is an open convex set,

$$C \subset U$$

is a nonempty closed and convex set, and $f : U \to R^1$ is a convex function that is Lipschitz on all bounded sets in U. We assume that f is Frechet differentiable at any point $x \in U$ such that $f(x) > \inf(f, C)$.

We assume that the following two assumptions hold:

(A1) There exists a nonempty bounded set $C_0 \subset C$ such that $\inf(f, C_0) = \inf(f, C)$.

(A2) For every pair $M, \epsilon > 0$, the Frechet derivative $f'(\cdot)$ is uniformly continuous on the set

$$\{x \in B(0, M) \cap C : f(x) > \inf(f, C) + \epsilon\}.$$

In this section, we consider the minimization problem

$$f(x) \to \min, \ x \in C.$$

Let $\lambda \in (0, 1]$ and $c_* \geq 1$ be fixed. We recall the following definition (see (2.1) and (2.2)).

For each $x \in C$ and each pair of numbers $\delta, \Delta \in (0, 1]$, denote by $E(x, \delta, \Delta)$ the set of all $l \in C - x$ for which there exist $g \in X^*$ such that

$$\|g - f'(x)\|_* \leq \delta$$

and

$$\xi \in B(0, c_*) \cap B(l, \delta)$$

such that

$$\langle g, \xi \rangle \leq \lambda \inf\{\langle g, \eta \rangle : \eta \in B(0, c_*) \cap (C - x)\} + \Delta.$$

Note that Lemma 2.1 holds and implies the following auxiliary result.

Lemma 4.4. *Let $\bar{K} \geq 1$, $\bar{\epsilon} \in (0, 1]$. Then there exist $\gamma, \Delta \in (0, 1)$ such that for each $x \in B(0, \bar{K}) \cap C$ satisfying*

$$f(x) \geq \inf(f, C) + \bar{\epsilon}$$

and each $l \in E(x, \Delta, \Delta)$, the inequality $\langle f'(x), l \rangle \leq -\gamma$ holds.

Proof. Lemma 2.1 implies that there exist $\bar{\alpha}, \gamma, \Delta \in (0, 1)$ such that the following property holds:

(a) For each $x \in B(0, \bar{K}) \cap C$ satisfying

$$f(x) \geq \inf(f, C) + \bar{\epsilon},$$

each $l \in E(x, \Delta, \Delta)$, and each $\beta \in (0, \bar{\alpha}]$, the inequality

$$f(x + \beta l) - f(x) \leq -\beta \gamma \tag{4.2}$$

holds.

Let $x \in B(0, \bar{K}) \cap C$ satisfy

$$f(x) \geq \inf(f, C) + \bar{\epsilon}$$

and $l \in E(x, \delta, \Delta)$. By property (a), for each $\beta \in (0, \bar{\alpha}]$, Eq. (4.2) holds and implies $\langle f'(x), l \rangle \leq -\gamma$. Lemma 4.4 is proved.

In this chapter, we prove the following results.

Theorem 4.5. *Assume that*

$$\lim_{\|x\| \to \infty, \, x \in U} f(x) = \infty \tag{4.3}$$

and $K > 0, \epsilon \in (0, 1)$. Then there exist $N_0 > 0$ and $\Delta > 0$ such that for each $T \geq N_0$ and each $x \in W^{1,1}(0, T)$ that satisfies

$$x(t) \in C, \; t \in [0, T], \tag{4.4}$$

$$\|x(0)\| \leq K, \tag{4.5}$$

and

$$x'(t) \in E(x(t), \Delta, \Delta), \ t \in [0, T] \ a. \ e.,$$ (4.6)

the inequality

$$f(x(t)) \le \inf(f, C) + \epsilon$$

holds for all $t \in [N_0, T]$.

Theorem 4.6. *Assume that* $x : [0, \infty) \to C$ *is a bounded function, for every* $T > 0$, $x \in W^{1,1}(0, T)$, $\Delta(t) > 0$ *for every* $t \ge 0$,

$$\lim_{t \to \infty} \Delta(t) = 0$$ (4.7)

and that

$$x'(t) \in E(x(t), \Delta(t), \Delta(t)), \ t \in [0, \infty) \ a. \ e.$$ (4.8)

Then

$$\lim_{t \to \infty} f(x(t)) = \inf(f, C).$$

Theorem 4.7. *Assume that* f *is Frechet differentiable at every point of* U, $x : [0, \infty) \to C$, *for every* $T > 0$, $x \in W^{1,1}(0, T)$,

$$\liminf_{t \to \infty} \|x(t)\| < \infty,$$ (4.9)

$\Delta(t) > 0$ *for every* $t \ge 0$,

$$\lim_{t \to \infty} \Delta(t) = 0,$$ (4.10)

and that for a. e. $t \in [0, T]$,

$$x'(t) \in E(x(t), \Delta(t), \Delta(t))$$ (4.11)

and

$$\langle f'(x(t), x'(t)) \rangle \le 0.$$ (4.12)

Then

$$\lim_{t \to \infty} f(x(t)) = \inf(f, C).$$

Theorems 4.5–4.7 are proved in Sects. 4.3–4.5, respectively.

4.3 Proof of Theorem 4.5

By (4.3), we may assume without loss of generality that

$$\{x \in C : \ f(x) \le \inf(f, C) + 4\} \subset B(0, K).$$ (4.13)

Fix

$$K_0 > \sup\{|f(x)| : \ x \in B(0, K) \cap C\}.$$ (4.14)

By (4.3), there exists $K_1 > K + 1$ such that

$$\{x \in C : f(x) \leq K_0 + 1\} \subset B(0, K_1 - 1). \tag{4.15}$$

Lemma 4.4 implies that there exist $\gamma, \Delta, \in (0, 1)$ such that the following property holds:

(a) For each $x \in B(0, K_1) \cap C$ satisfying

$$f(x) \geq \inf(f, C) + \epsilon/4$$

and each $l \in E(x, \Delta, \Delta)$, the inequality $\langle f'(x), l \rangle \leq -\gamma$ holds.

Choose

$$N_0 > \gamma^{-1}(K_0 - \inf(f, C)).$$

Assume that $T \geq N_0$, $x \in W^{1,1}(0, T)$, and (4.4)–(4.6) hold. In view of (4.6),

$$\|x'(t)\| \leq c_* + 1 \leq 2c_*, \ t \in [0, T] \text{ a. e.} \tag{4.16}$$

By (4.16), x and $f \circ x$ are Lipschitz on $[0, T]$. Assume that $\tau \in [0, T]$,

$$[\tau, \tau + N_0] \subset [0, T], \tag{4.17}$$

$$\|x(\tau)\| \leq K. \tag{4.18}$$

We show that there exists $s \in [\tau, \tau + N_0]$ such that

$$f(x(s)) < \inf(f, C) + \epsilon/4.$$

Assume the contrary. Then

$$f(x(s)) \geq \inf(f, C) + \epsilon/4, \ s \in [\tau, \tau + N_0]. \tag{4.19}$$

We show that

$$\|x(t)\| \leq K_1, \ t \in [\tau, \tau + N_0].$$

Assume the contrary. Then there exists

$$\tilde{t} \in [\tau, \tau + N_0]$$

such that

$$\|x(\tilde{t})\| > K_1.$$

Together with (4.18) and the inequality $K_1 > K$, this implies that there exists

$$\tau_0 \in [\tau, \tau + N_0]$$

such that

$$\tau_0 > \tau, \tag{4.20}$$

$$\|x(\tau_0)\| = K_1, \ \|x(t)\| < K_1, \ t \in [\tau, \tau_0). \tag{4.21}$$

Property (a) and Eqs. (4.6) and (4.19)–(4.21) imply that for a. e. $t \in [\tau, \tau_0)$,

$$\langle f'(x(t)), x'(t) \rangle \le -\gamma. \tag{4.22}$$

Lemma 4.3, (4.14), and (4.22) imply that

$$f(x(\tau_0)) - f(x(\tau)) = \int_\tau^{\tau_0} \langle f'(x(t)), x'(t) \rangle dt \le 0. \tag{4.23}$$

By (4.14), (4.18), and (4.23),

$$f(x(\tau_0)) \le f(x(\tau)) \le K_0,$$

$$\|x(\tau_0)\| \le K_1 - 1.$$

This contradicts (4.21). The contradiction we have reached proves that

$$\|x(t)\| \le K_1, \ t \in [\tau, \tau + N_0]. \tag{4.24}$$

Property (a) and Eqs. (4.6), (4.19) and (4.24) imply that for a. e. $t \in [\tau, \tau+N_0]$,

$$\langle f'(x(t)), x'(t) \rangle dt \le -\gamma. \tag{4.25}$$

By (4.3), (4.19), and (4.25),

$$f(x(\tau + N_0)) - f(x(\tau)) = \int_\tau^{\tau+N_0} \langle f'(x(t)), x'(t) \rangle dt \le -\gamma N_0. \tag{4.26}$$

It follows from (4.14), (4.18), and (4.26) that

$$\inf(f, C) \le f(x(\tau + N_0)) \le K_0 - \gamma N_0$$

and

$$N_0 \le \gamma^{-1}(K_0 - \inf(f, C)).$$

This contradicts the choice of N_0. The contradiction we have reached proves that there exists $s \in [\tau, \tau + N_0]$ such that

$$f(x(s)) < \inf(f, C) + \epsilon/4.$$

Together with (4.5), this implies that there exists

$$\tau_0 \in [0, N_0] \tag{4.26}$$

such that

$$f(x(\tau_0)) \le \inf(f, C) + \epsilon/4. \tag{4.27}$$

We show that for all $t \in [\tau_0, T]$, we have

$$f(x(t)) \leq \inf(f, C) + \epsilon.$$

Assume the contrary. Then by (4.27), there exists

$$\tau_1 \in (\tau_0, T]$$

such that

$$f(x(\tau_1)) = \inf(f, C) + \epsilon, \tag{4.28}$$

and for every $t \in [\tau_0, \tau_1)$,

$$f(x(t)) < \inf(f, C) + \epsilon. \tag{4.29}$$

By (4.13), (4.28), and (4.29),

$$\|x(t)\| \leq K, \ t \in [\tau_0, \tau_1]. \tag{4.30}$$

In view of (4.28), there exists $\tau_2 \in (\tau_0, \tau_1)$ such that

$$f(x(t)) \geq \inf(f, C) + \epsilon/2, \ t \in [\tau_2, \tau_1]. \tag{4.31}$$

Property (a), (4.6), (4.30), and (4.31) imply that

$$\langle f'(x(t)), x'(t) \rangle dt \leq -\gamma, \ t \in [\tau_2, \tau_1] \text{ a. e.} \tag{4.32}$$

Lemma 4.3 and Eqs. (4.29), (4.31), and (4.32) imply that the function $f(x(\cdot))$ is decreasing on $[\tau_2, \tau_1]$ and

$$f(x(\tau_1)) \leq f(x(\tau_2)) < \inf(f, C) + \epsilon.$$

This contradicts (4.28). The contradiction we have reached proves that

$$f(x(t)) \leq \inf(f, C) + \epsilon, \ t \in [\tau_0, T]$$

and completes the proof of Theorem 4.5.

4.4 Proof of Theorem 4.6

There exists $K > 0$ such that

$$\|x(t)\| \leq K, \ \in [0, \infty). \tag{4.33}$$

Let $\epsilon \in (0, 1)$. It is sufficient to show that

$$f(x(t)) \leq \inf(f, C) + \epsilon \text{ for all sufficiently large numbers } t > 0.$$

Lemma 4.4 implies that there exist $\gamma, \Delta, \in (0, 1)$ such that the following property holds:

(a) For each $x \in B(0, K) \cap C$ satisfying

$$f(x) \geq \inf(f, C) + \epsilon/4$$

and each $l \in E(x, \Delta, \Delta)$, the inequality $\langle f'(x(t), l) \rangle \leq -\gamma$ holds.

In view of (4.7), there exists $\tau_0 > 0$ such that for each $t \geq \tau_0$,

$$\Delta(t) < \Delta. \tag{4.34}$$

We show that there exists $s > \tau_0$ such that

$$f(x(s)) \leq \inf(f, C) + \epsilon/4.$$

Assume the contrary. Then for each $s > \tau_0$,

$$f(x(s)) > \inf(f, C) + \epsilon/4. \tag{4.35}$$

Property (a) and Eqs. (4.8) and (4.33)–(4.35) imply that for a. e. $t > \tau_0$,

$$\langle f'(x(t)), x'(t) \rangle \leq -\gamma. \tag{4.36}$$

Clearly, the functions x and $f \circ x$ are Lipschitz on $[0, \infty)$. Lemma 4.3 and Eqs. (4.33), (4.35), and (4.36) imply that

$$\inf(f, C) - f(x(\tau_0)) \leq f(x(\tau)) - f(x(\tau_0))$$

$$= \int_{\tau_0}^{\tau} \langle f'(x(t)), x'(t) \rangle dt \leq -\gamma(\tau - \tau_0) \to -\infty$$

as $\tau \to \infty$. The contradiction we have reached proves that there exists $s_0 > \tau_0$ such that

$$f(x(s_0)) \leq \inf(f, C) + \epsilon/4. \tag{4.37}$$

We show that for all $t > s_0$, we have

$$f(x(t)) \leq \inf(f, C) + \epsilon.$$

Assume the contrary. Then by (4.37), there exists

$$s_1 > s_0$$

such that

$$f(x(s_1)) = \inf(f, C) + \epsilon/2, \tag{4.38}$$

$$f(x(t)) < \inf(f, C) + \epsilon/2, \ t \in [s_0, s_1). \tag{4.39}$$

In view of (4.37) and (4.38), there exists $s_2 \in (s_0, s_1]$ such that

$$f(x(t)) \geq \inf(f, C) + \epsilon/4, \ t \in [s_2, s_1]. \tag{4.40}$$

Property (a), (4.8), (4.33), (4.34), and (4.40) imply that

$$\langle f'(x(t)), x'(t) \rangle < -\gamma, \ t \in [s_2, s_1] \text{ a. e.} \tag{4.41}$$

Lemma 4.3 and Eqs. (4.38)–(4.41) imply that

$$\inf(f, C) + \epsilon/2 = f(x(s_1))$$

$$= f(x(s_2)) + \int_{s_2}^{s_1} \langle f'(x(t)), x'(t) \rangle dt \leq \inf(f, C) + \epsilon/2 - \gamma(s_1 - s_2),$$

a contradiction. The contradiction we have reached proves that

$$f(x(t)) \leq \inf(f, C) + \epsilon, \ t \in (s_0, \infty)$$

and completes the proof of Theorem 4.6.

4.5 Proof of Theorem 4.7

Clearly, x and $f \circ x$ are Lipschitz on $[0, \infty)$. Lemma 4.3 and (4.12) imply that the function $f \circ x$ is decreasing on $[0, \infty)$. We show that

$$\lim_{t \to \infty} f(x(t)) = \inf(f, C).$$

Assume the contrary. Then there exists $\epsilon \in (0, 1)$ such that

$$f(x(t)) \geq \inf(f, C) + \epsilon, \ t \in [0, \infty). \tag{4.42}$$

There exists $K > 0$ such that

$$\liminf_{t \in \infty} \|x(t)\| < K. \tag{4.43}$$

There exists a sequence of positive numbers $\{t_i\}_{i=1}^{\infty}$ such that for each integer $i \geq 1$,

$$t_{i+1} > t_i + 2, \ \|x(t_i)\| < K. \tag{4.44}$$

In view of (4.11), for a. e. $t \in (0, \infty)$,

$$\|x'(t)\| \leq c_* + 1 \leq 2c_*. \tag{4.45}$$

By (4.45), for each integer $i \geq 0$ and each $t \in [t_i, t_i + 2]$,

$$\|x(t)\| \leq \|x(t_i)\| + 2c_*(t - t_i) \leq K + 2c_*. \tag{4.46}$$

Lemma 4.4 implies that there exist $\gamma, \Delta, \in (0, 1)$ such that the following property holds:

(a) For each $x \in B(0, K + 2c_*) \cap C$ satisfying

$$f(x) \geq \inf(f, C) + \epsilon/2$$

and each $l \in E(x, \Delta, \Delta)$, the inequality $\langle f'(x(t)), l \rangle \leq -\gamma$ holds.

By (4.10), there exists an integer $k_0 \geq 1$ such that

$$\Delta(t) \leq \Delta, \ t \in [t_{k_0}, \infty). \tag{4.47}$$

Property (a) and Eqs. (4.11), (4.42), (4.46), and (4.47) imply that for a. e. $t \in [t_k, t_k + 1]$,

$$\langle f'(x(t)), x'(t) \rangle \leq -\gamma. \tag{4.48}$$

Lemma 4.3 and Eqs. (4.44) and (4.48) imply that for every integer $k \geq k_0$,

$$f(x(t_k + 1)) - f(x(t_k)) = \int_{t_k}^{t_k+1} \langle f'(x(t)), x'(t) \rangle dt \leq -\gamma$$

and

$$f(x(t_{k+1})) \leq f(x(t_k + 1)) \leq f(x(t_k)) - \gamma.$$

This implies that for each integer $i \geq 0$,

$$\inf(f, C) \leq f(x_{t_{k_0}+i}) \leq f(x_{t_{k_0}}) - i\gamma \to -\infty$$

as $i \to \infty$, a contradiction. The contradiction we have reached proves that

$$\lim_{t \to \infty} f(x(t)) = \inf(f, C)$$

and completes the proof of Theorem 4.7.

4.6 The First Convergence Result with Estimations

Assume that $(X, \| \cdot \|)$ is a Banach space, $U \subset X$ is an open convex set,

$$C \subset U$$

is a nonempty closed and convex set, and $f : U \to R^1$ is a convex function that is Frechet differentiable at any point $x \in U$.

In this section, we consider the minimization problem

$$f(x) \to \min, \ x \in C.$$

Let $K \geq 1$, $\lambda \in (0, 1]$, and $c_* > 1$ be fixed.
Assume that

$$\inf(f, C) = \inf(f, C \cap B(0, K)), \tag{4.49}$$

$$K_0 \geq \sup\{|f(x)| : x \in C \cap B(0, K)\} + 1, \tag{4.50}$$

$$K_1 > \sup\{\|x\| : x \in U, |f(x)| \leq K_0 + 1\}, \tag{4.51}$$

$L_0 \geq 1$, $\rho \in (0, 1]$ and that

$$|f(z_1) - f(z_2)| \leq L_0 \|z_1 - z_2\|, \ z_1, z_2 \in B(0, K_1 + 1) \cap U, \tag{4.52}$$

$$\|f'(z_1) - f'(z_2)\|_* \leq L_0 \|z_1 - z_2\|^\rho, \ z_1, z_2 \in B(0, K_1 + 1) \cap U. \tag{4.53}$$

Let $\epsilon \in (0, 1]$,

$$\Delta = (64K_1)^{-1}\epsilon\lambda, \ \gamma = (64K_1)^{-1}\epsilon\lambda, \tag{4.54}$$

$$\delta = (64K_1)^{-1}\epsilon\lambda(2c_* + L_0)^{-1}, \tag{4.55}$$

$$\alpha = \min\{((2^6 K_1 c_*(L_0 + 1))^{-1}\epsilon\lambda)^{1/\rho}, (8c_* L_0)^{-1}\epsilon\}. \tag{4.56}$$

For each $x \in C$ and each pair of numbers $\delta, \Delta \in (0, 1]$, we consider the set $E(x, \delta, \Delta)$ defined in Sect. 4.2 (see also Sect. 2.13, (2.238)–(2.240)). Note that Lemma 2.15 holds and implies the following auxiliary result.

Lemma 4.8. *Let* $x \in B(0, K_1) \cap C$ *satisfy*

$$f(x) \geq \inf(f, C) + \bar{\epsilon}/4$$

and $l \in E(x, \delta, \Delta)$. *Then* $\langle f'(x), l \rangle \leq -\gamma$.

Proof. Lemma 2.15 implies that for all $\beta \in (0, \alpha]$,

$$\beta^{-1}(f(x + \beta l) - f(x)) \leq -\gamma.$$

This implies that $\langle f'(x), l \rangle \leq -\gamma$. Lemma 4.8 is proved.

Theorem 4.9. *Let*

$$N_0 > 2K_0\gamma^{-1}. \tag{4.57}$$

Assume that $T \geq N_0$, $x \in W^{1,1}(0, T)$ *satisfies*

$$x(t) \in C, \ t \in [0, T], \tag{4.58}$$

$$\|x(0)\| \leq K, \tag{4.59}$$

and

$$x'(t) \in E(x(t), \delta, \Delta), \ t \in [0, T] \ a. \ e. \tag{4.60}$$

Then the inequality

$$f(x(t)) \leq \inf(f, C) + \epsilon$$

holds for all $t \in [N_0, T]$.

Proof. By (4.55) and (4.60),

$$\|x'(t)\| \le c_* + 1 \le 2c_*, \ t \in [0,T] \text{ a. e.} \tag{4.61}$$

Assume that $\tau \in [0,T]$ and

$$[\tau, \tau + N_0] \subset [0,T], \tag{4.62}$$

$$f(x(\tau)) < K_0. \tag{4.63}$$

We show that there exists $s \in [\tau, \tau + N_0]$ such that

$$f(x(s)) < \inf(f, C) + \epsilon/4.$$

Assume the contrary. Then

$$f(x(s)) \ge \inf(f, C) + \epsilon/4, \ s \in [\tau, \tau + N_0]. \tag{4.64}$$

We show that

$$f(x(t)) \le K_0, \ t \in [\tau, \tau + N_0].$$

Assume the contrary. Then there exists

$$\tilde{t} \in [\tau, \tau + N_0] \tag{4.65}$$

such that

$$f(x(\tilde{t})) > K_0. \tag{4.66}$$

By (4.63) and (4.66), there exists

$$\tilde{\tau} \in (\tau, \tilde{t}]$$

such that

$$f(x(t)) \le K_0, \ t \in [\tau, \tilde{\tau}], \ f(x(\tilde{\tau})) = K_0. \tag{4.67}$$

In view of (4.51) and (4.67), for all $t \in [\tau, \tilde{\tau}]$,

$$\|x(t)\| \le K_1. \tag{4.68}$$

Lemma 4.8 and Eqs. (4.64)–(4.66) and (4.68) imply that for a. e. $t \in [\tau, \tilde{\tau}]$,

$$\langle f'(x(t)), x'(t) \rangle \le -\gamma. \tag{4.69}$$

In view of (4.61), x is Lipschitz on the interval $[0,T]$. Together with Lemma 4.3 and (4.69), this implies that

$$f(x(\tilde{\tau})) - f(x(\tau)) = \int_\tau^{\tilde{\tau}} \langle f'(x(t)), x'(t) \rangle dt < 0.$$

Combined with (4.63), this implies that

$$f(x(\tilde{\tau})) \leq f(x(\tau)) < K_0.$$

This contradicts (4.67). The contradiction we have reached proves that

$$f(x(t)) \leq K_0, \ t \in [\tau, \tau + N_0]. \tag{4.70}$$

By (4.51) and (4.70),

$$\|x(t)\| \leq K_1, \ t \in [\tau, \tau + N_0]. \tag{4.71}$$

Lemma 4.8 and Eqs. (4.60), (4.64) and (4.71) imply that for a. e. $t \in [\tau, \tau + N_0]$,

$$\langle f'(x(t)), x'(t) \rangle \leq -\gamma. \tag{4.72}$$

Lemma 4.3 and (4.72) imply that

$$f(x(\tau + N_0)) - f(x(\tau)) = \int_{\tau}^{\tau+N_0} \langle f'(x(t)), x'(t) \rangle dt \leq -\gamma N_0.$$

Combined with (4.49), (4.50), and (4.63), this implies that

$$\gamma N_0 \leq f(x(\tau)) - f(x(\tau + N_0)) \leq K_0 - \inf(f, C) \leq 2K_0$$

and

$$N_0 \leq 2\gamma^{-1} K_0.$$

This contradicts (4.57). The contradiction we have reached proves that there exists $s \in [\tau, \tau + N_0]$ such that

$$f(x(s)) \leq \inf(f, C) + \epsilon/4.$$

Equations (4.50) and (4.59) imply that

$$f(x(0)) < K_0.$$

Therefore, there exists

$$\tau_0 \in [0, N_0] \tag{4.73}$$

such that

$$f(x(\tau_0)) \leq \inf(f, C) + \epsilon/4. \tag{4.74}$$

We show that for all $t \in [\tau_0, T]$, we have

$$f(x(t)) \leq \inf(f, C) + \epsilon.$$

Assume the contrary. Then there exists

$$\tau_1 \in (\tau_0, T]$$

such that

$$f(x(\tau_1)) = \inf(f, C) + \epsilon, \tag{4.75}$$

$$f(x(t)) < \inf(f, C) + \epsilon, \ t \in [\tau_0, \tau_1). \tag{4.76}$$

By (4.59)–(4.61) and (4.76),

$$\|x(t)\| \le K_1, \ t \in [\tau_0, \tau_1]. \tag{4.77}$$

In view of (4.75) and (4.76), there exists $\tau_2 \in (\tau_0, \tau_1)$ such that

$$f(x(t)) \ge \inf(f, C) + \epsilon/2, \ t \in [\tau_2, \tau_1]. \tag{4.78}$$

Lemma 4.8 and Eqs. (4.60), (4.77), and (4.78) imply that

$$\langle f'(x(t)), x'(t) \rangle dt \le -\gamma, \ t \in [\tau_2, \tau_1] \ \text{a. e.} \ . \tag{4.79}$$

Lemma 4.3 and Eq. (4.79) imply that the function $f(x(\cdot))$ is strictly decreasing on $[\tau_2, \tau_1]$ and

$$f(x(\tau_2)) > f(x(\tau_1)) = \inf(f, C) + \epsilon.$$

This contradicts (4.76). The contradiction we have reached completes the proof of Theorem 4.9.

4.7 Nonconvex Optimization

Assume that $(X, \| \cdot \|)$ is a Banach space, $U \subset X$ is an open convex set,

$$C \subset U$$

is a nonempty closed and convex set, and a function $f : U \to R^1$ is Frechet differentiable at any point $x \in U$ and is Lipschitz on all bounded sets in U.

Recall (see (3.1)) that for each $x \in C$,

$$\Lambda(x) = \inf\{\langle f'(x), h \rangle : \ h \in (C - x) \cap B(0, 1)\}. \tag{4.80}$$

We assume that the following assumption holds:

(A1) For every $M > 0$, the Frechet derivative $f'(\cdot)$ is uniformly continuous on the set $B(0, M) \cap C$.

We consider the minimization problem

$$f(x) \to \min, \ x \in C.$$

Let $c_* > 1$ and $\lambda \in (0, 1]$ be fixed.

Recall (see (3.2)–(3.4)) that for each $x \in C$ and each pair of numbers $\delta, \Delta \in (0, 1]$, $E(x, \delta, \Delta)$ is the set of all $l \in C - x$ for which there exist $g \in X^*$ such that

$$\|g - f'(x)\|_* \le \delta \tag{4.81}$$

and

$$\xi \in B(0, c_*) \cap B(l, \delta) \tag{4.82}$$

such that

$$\langle g, \xi \rangle \leq \lambda \inf\{\langle g, \eta \rangle : \eta \in B(0, c_*) \cap (C - x)\} + \Delta. \tag{4.83}$$

Note that Lemmas 3.1, 3.3, and 3.4 and Proposition 3.2 hold. Lemma 3.4 implies the following result.

Lemma 4.10. *Let* $\bar{K} \geq 1$, $\bar{\epsilon} \in (0, 1)$. *Then there exist* $\gamma, \delta, \Delta \in (0, 1)$ *such that for each* $x \in B(0, \bar{K}) \cap C$ *satisfying*

$$\Lambda(x) \leq -\bar{\epsilon}$$

and each $l \in E(x, \delta, \Delta)$, *the inequality* $\langle f'(x), l \rangle \leq -\gamma$ *holds.*

Proof. Lemma 3.4 implies that there exist $\alpha, \gamma, \Delta, \delta \in (0, 1)$ such that the following property holds:

(a) For each $x \in B(0, \bar{K}) \cap C$ satisfying $\Lambda(x) \leq -\bar{\epsilon}$, each $l \in E(x, \delta, \Delta)$, and each $\beta \in (0, \alpha]$, the inequality

$$f(x) - f(x + \beta l) \geq \beta\gamma$$

holds.

Assume that $x \in B(0, \bar{K}) \cap C$ satisfies

$$\Lambda(x) \leq -\bar{\epsilon}$$

and $l \in E(x, \delta, \Delta)$. Property (a) implies that for each $\beta \in (0, \alpha)$,

$$\beta^{-1}(f(x + \beta l) - f(x)) \leq -\gamma.$$

This implies that $\langle f'(x), l \rangle \leq -\gamma$. Lemma 4.10 is proved.

4.8 Convergence Results

In this chapter, we establish the following convergence results. The first one is proved in Sect. 4.9.

Theorem 4.11. *Assume that*

$$\lim_{\|x\| \to \infty, \, x \in C} f(x) = \infty$$

and $K > 0, \epsilon \in (0, 1]$. *Then there exist* $N_0 > 0$ *and* $\Delta > 0$ *such that for each* $x \in W^{1,1}(0, N_0)$ *that satisfies*

$$\|x(0)\| \leq K,$$

$$x(t) \in C, \ t \in [0, N_0],$$

and

$$x'(t) \in E(x(t), \Delta, \Delta), \ t \in [0, N_0] \ a. \ e.,$$

the inequality

$$\sup\{\Lambda(x(t)) : \ t \in [0, N_0]\} \geq -\epsilon$$

holds.

Our second result is proved in Sect. 4.10.

Theorem 4.12. *Assume that*

$$\lim_{\|x\| \to \infty, \ x \in C} f(x) = \infty$$

and $K > 0, \epsilon \in (0, 1)$. Then there exist $N_0 > 0$ and $\Delta > 0$ such that for each $T \geq N_0$ and each $x \in W^{1,1}(0, T)$ that satisfies

$$\|x(0)\| \leq K,$$

$$x(t) \in C, \ t \in [0, T],$$

$$x'(t) \in E(x(t), \Delta, \Delta), \ t \in [0, T] \ a. \ e.,$$

and

$$\langle f'(x(t)), x'(t) \rangle \leq 0, \ t \in [0, T] \ a. \ e.,$$

the inequality

$$mes \left(\{ t \in [0, T] : \ \Lambda(x(t)) \leq -\epsilon \} \right) \leq N_0$$

holds.

The next result follows from Theorem 4.11.

Theorem 4.13. *Assume that the set C is bounded and $\epsilon \in (0, 1)$. Then there exists $\Delta \in (0, 1)$ such that for each $x : [0, \infty) \to C$ satisfying $x \in W^{1,1}(0, T)$ for every $T > 0$ and such that for a. e. $t \in [0, \infty)$,*

$$x'(t) \in E(x(t), \Delta(t), \Delta(t)),$$

the inequality

$$\limsup_{t \to \infty} \Lambda(x(t)) \geq -\epsilon$$

holds.

The following result is proved in Sect. 4.10.

Theorem 4.14. *Assume that $x : [0, \infty) \to C$ is bounded, for every $T > 0$, $x \in W^{1,1}(0, T)$, $\Delta(t) \in (0, 1)$ for every $t \geq 0$,*

$$\lim_{t \to \infty} \Delta(t) = 0$$

and that for a. e. $t \in [0, \infty)$,

$$x'(t) \in E(x(t), \Delta(t), \Delta(t)). \tag{4.84}$$

Then

$$\limsup_{t \to \infty} \Lambda(x(t)) \geq 0.$$

The next result is proved in Sect. 4.11.

Theorem 4.15. *Assume that $K > 0, \epsilon > 0$. Then there exists $\Delta > 0$ such that for each $x : [0, \infty) \to C$ satisfying for every $T > 0$, $x \in W^{1,1}(0, T)$,*

$$\|x(t)\| \leq K, \ t \in [0, \infty),$$

$$x'(t) \in E(x(t), \Delta, \Delta), \ t \in [0, \infty) \ a. \ e.$$

and

$$\langle f'(x(t)), x'(t) \rangle \leq 0, \ t \in [0, \infty) \ a. \ e.,$$

the inequality

$$\Lambda(x(t)) \geq -\epsilon$$

holds for all sufficiently large $t > 0$.

The next result follows from Theorem 4.15.

Theorem 4.16. *Assume that $x : [0, \infty) \to C$ is bounded, for every $T > 0$, $x \in W^{1,1}(0, T)$, $\Delta(t) \in (0, 1)$ for every $t \geq 0$,*

$$\lim_{t \to \infty} \Delta(t) = 0$$

and that for a. e. $t \in [0, \infty)$,

$$x'(t) \in E(x(t), \Delta(t), \Delta(t)),$$

$$\langle f'(x(t)), x'(t) \rangle \leq 0, \ t \in [0, \infty) \ a. \ e.$$

Then

$$\lim_{t \to \infty} \Lambda(x(t)) = 0.$$

The following result is proved in Sect. 4.12.

Theorem 4.17. *Assume that $K > 0, \epsilon \in (0,1]$. Then there exist $K_1 > 0$ and $\Delta \in (0,1)$ such that for each $T > 0$ and each $x \in W^{1,1}(0,T)$ that satisfies*

$$x(t) \in C, \ t \in [0,T],$$

$$\|x(t)\| \leq K, \ t \in [0,T]$$

$$x'(t) \in E(x(t), \Delta, \Delta), \ t \in [0,T] \ a. \ e.,$$

the inequality

$$mes \ (\{t \in [0,T] : \ \Lambda(x(t)) \leq -\epsilon\}) \leq \epsilon T + K_1$$

holds.

Theorem 4.17 implies the next result.

Theorem 4.18. *Assume that $x : [0,\infty) \to C$ is bounded, for every $T > 0$, $x \in W^{1,1}(0,T)$, $\Delta(t) \in (0,1)$ for every $t \geq 0$,*

$$\lim_{t \to \infty} \Delta(t) = 0,$$

and that for a. e. $t \in [0,\infty)$,

$$x'(t) \in E(x(t), \Delta(t), \Delta(t)).$$

Then for each $\epsilon > 0$,

$$T^{-1} mes \ (\{t \in [0,T] : \ \Lambda(x(t)) \leq -\epsilon\}) \leq \epsilon$$

for all sufficiently large $T > 0$.

4.9 Proof of Theorem 4.11

We may assume without loss of generality that

$$\inf\{f(x) : \ x \in B(0,K) \cap C\} = \inf(f,C). \tag{4.85}$$

Fix

$$K_0 > \sup\{|f(x)| : \ x \in B(0,K) \cap C\} + 4 \tag{4.86}$$

and $K_1 > K + 4$ such that

$$\{x \in C : \ f(x) \leq K_0 + 2\} \subset B(0,K_1). \tag{4.87}$$

Lemma 4.10 implies that there exist $\gamma, \Delta, \in (0,1)$ such that the following property holds:

(a) For each $x \in B(0, K_1) \cap C$ satisfying $\Lambda(x) \leq -\epsilon$ and each $l \in E(x, \Delta, \Delta)$, the inequality $\langle f'(x(t), l) \leq -\gamma$ holds.

Choose
$$N_0 > 2\gamma^{-1} K_0.$$

Assume that $x \in W^{1,1}(0, N_0)$,

$$\|x(0)\| \leq K, \tag{4.88}$$

$$x(t) \in C, \ t \in [0, N_0] \tag{4.89}$$

and
$$x'(t) \in E(x(t), \Delta, \Delta), \ t \in [0, N_0] \text{ a. e.} \tag{4.90}$$

We show that there exists $s \in [0, N_0]$ such that

$$\Lambda(x(s)) \geq -\epsilon.$$

Assume the contrary. Then

$$\Lambda(x(t)) < -\epsilon, \ t \in [0, N_0]. \tag{4.91}$$

We show that
$$f(x(t)) \leq K_0, \ t \in [0, N_0].$$

Assume the contrary. Then in view of (4.86) and (4.88), there exists

$$s \in (0, N_0) \tag{4.92}$$

such that
$$f(x(t)) \leq K_0, \ t \in [0, s], \ f(x(s)) = K_0. \tag{4.93}$$

By (4.87) and (4.93),
$$\|x(t)\| \leq K_1, \ t \in [0, s]. \tag{4.94}$$

Property (a) and Eqs. (4.90), (4.91), (4.93), and (4.94) imply that for a. e. $t \in [0, s]$,
$$\langle f'(x(t)), x'(t) \rangle \leq -\gamma. \tag{4.95}$$

In view of (4.90), x is Lipschitz on $[0, N_0]$. Lemma 4.3 and (4.95) imply that

$$f(x(s)) - f(x(0)) = \int_0^s \langle f'(x(t)), x'(t) \rangle dt \leq -\gamma s$$

and
$$f(x(s)) \leq f(x(0)) \leq K_0.$$

This contradicts (4.93). The contradiction we have reached proves that

$$f(x(t)) \leq K_0, \ t \in [0, N_0]. \tag{4.96}$$

By (4.87) and (4.96),
$$\|x(t)\| \leq K_1, \ t \in [0, N_0]. \tag{4.97}$$
Property (a) and Eqs. (4.90), (4.91), and (4.97) imply that for a. e. $t \in [0, N_0]$,
$$\langle f'(x(t)), x'(t) \rangle \leq -\gamma. \tag{4.98}$$
Lemma 4.3 and Eqs. (4.85), (4.86), (4.96), and (4.98) imply that
$$f(x(N_0)) - f(x(0)) = \int_0^{N_0} \langle f'(x(t)), x'(t) \rangle dt \leq -\gamma N_0,$$

$$\gamma N_0 \leq f(x(0)) - f(x(N_0)) \leq K_0 - \inf(f, C) \leq 2K_0,$$
and
$$N_0 \leq 2\gamma^{-1} K_0.$$
This contradicts the definition of N_0. The contradiction we have reached proves that there exists $s \in [0, N_0]$ such that
$$\Lambda(x(s)) \geq -\epsilon.$$
This completes the proof of Theorem 4.11.

4.10 Proofs of Theorems 4.12 and 4.14

Proof of Theorem 4.12 We may assume without loss of generality that
$$\inf\{f(x) : \ x \in B(0, K) \cap C\} = \inf(f, C). \tag{4.99}$$
Fix
$$K_0 > \sup\{|f(x)| : \ x \in B(0, K + 1) \cap C\} \tag{4.100}$$
and $K_1 > K$ such that
$$\{x \in C : \ f(x) \leq K_0 + 1\} \subset B(0, K_1). \tag{4.101}$$
Lemma 4.10 implies that there exist $\gamma, \Delta, \in (0, 1)$ such that the following property holds:

(a) For each $x \in B(0, K_1) \cap C$ satisfying $\Lambda(x) \leq -\epsilon$ and each $l \in E(x, \Delta, \Delta)$, the inequality $\langle f'(x(t)), l \rangle \leq -\gamma$ holds.

Choose
$$N_0 > 2\gamma^{-1} K_0. \tag{4.102}$$
Assume that $T \geq N_0$, $x \in W^{1,1}(0, T)$,
$$\|x(0)\| \leq K, \tag{4.103}$$

$$x(t) \in C, \ t \in [0, T], \tag{4.104}$$

$$x'(t) \in E(x(t), \Delta, \Delta), \ t \in [0, T] \text{ a. e.}, \tag{4.105}$$

and

$$\langle f'(x(t)), x'(t) \rangle \leq 0, \ t \in [0, T] \text{ a. e.} \tag{4.106}$$

In view of (4.105), x is Lipschitz on $[0, T]$. Lemma 4.3 and (4.100) imply that the function $f(x(t))$, $t \in [0, T]$ is decreasing. Together with (4.100), (4.101), and (4.103), this implies that

$$f(x(t)) \leq K_0, \ t \in [0, T], \ \|x(t)\| \leq K_1, \ t \in [0, T]. \tag{4.107}$$

Lemma 4.3, property (a), and Eqs. (4.99), (4.100), (4.104), (4.106), and (4.107) imply that

$$-2K_0 \leq f(x(T)) - f(x(0)) = \int_0^T \langle f'(x(t)), x'(t) \rangle dt$$

$$\leq -\gamma \text{ mes } (\{t \in [0, T] : \ \Lambda(x(t)) \leq -\epsilon\})$$

and

$$\text{mes } (\{t \in [0, T] : \ \Lambda(x(t)) \leq -\epsilon\}) \leq 2K_0 \gamma^{-1} < N_0.$$

This completes the proof of Theorem 4.12.

Proof of Theorem 4.14 There exists $K > 0$ such that

$$\|x(t)\| \leq K, \ t \in [0, \infty). \tag{4.108}$$

Let $\epsilon \in (0, 1)$. Lemma 4.10 implies that there exist $\gamma, \Delta, \in (0, 1)$ such that the following property holds:

(a) For each $x \in B(0, K) \cap C$ satisfying $\Lambda(x) \leq -\epsilon$ and each $l \in E(x, \Delta, \Delta)$, the inequality $\langle f'(x(t)), l \rangle \leq -\gamma$ holds.

In order to complete the proof of the theorem, it is sufficient to show that

$$\limsup_{t \to \infty} \Lambda(x(t)) \geq -\epsilon.$$

Assume that the inequality above does not hold. Then there exists $t_0 > 0$ such that

$$\Lambda(x(t)) < -\epsilon, \ t \in [t_0, \infty). \tag{4.109}$$

There exists $t_1 > t_0$ such that

$$\Delta(t) \leq \Delta \text{ for each } t \geq t_1. \tag{4.110}$$

Property (a) and Eqs. (4.84), (4.108), and (4.109) imply that for a. e. $t \in [t_1, \infty)$,

$$\langle f'(x(t)), x'(t) \rangle \leq -\gamma. \tag{4.111}$$

Clearly, x is Lipschitz on $[0, \infty)$. Lemma 4.3 and Eq. (4.111) imply that

$$f(x(T)) - f(x(t_1)) = \int_{t_1}^T \langle f'(x(t)), x'(t) \rangle dt \leq -\gamma(T - t_1) \to -\infty$$

as $T \to \infty$ The contradiction we have reached proves Theorem 4.14.

4.11 Proof of Theorem 4.15

There exists $K_0 > 0$ such that

$$|f(x)| \leq K_0, \ x \in B(0, K) \cap C. \tag{4.112}$$

Lemma 4.10 implies that there exist $\gamma, \Delta, \in (0, 1)$ such that the following property holds:

(a) For each $x \in B(0, K) \cap C$ satisfying $\Lambda(x) \leq -\epsilon/2$ and each $l \in E(x, \Delta, \Delta)$, the inequality $\langle f'(x(t), l) \leq -\gamma$ holds.

Proposition 3.2 implies that there exists $\delta \in (0, 1)$ such that for each $z_1, z_2 \in B(0, K)$ satisfying $\|z_1 - z_2\| \leq \delta$, we have

$$|\Lambda(z_1) - \Lambda(z_2)| \leq \epsilon/4. \tag{4.113}$$

Assume that $x : [0, \infty) \to C$, $x \in W^{1,1}(0, T)$ for every $T > 0$,

$$\|x(t)\| \leq K, \ t \in [0, \infty), \tag{4.114}$$

$$x'(t) \in E(x(t), \Delta, \Delta), \ t \in [0, \infty) \text{ a. e.}, \tag{4.115}$$

and

$$\langle f'(x(t)), x'(t) \rangle \leq 0, \ t \in [0, \infty) \text{ a. e.} \tag{4.116}$$

Lemma 4.3 and (4.116) imply that the function $f \circ x$ is decreasing on $[0, \infty)$. Let $T > 0$ and

$$E_T = \{t \in [0, T] : \Lambda(x(t)) \leq -\epsilon/2\}. \tag{4.117}$$

Lemma 4.3, property (a), and Eqs. (4.112) and (4.114)–(4.117) imply that

$$\inf(f, C) - K_0 \leq f(x(T)) - f(x(0)) = \int_0^T \langle f'(x(t)), x'(t) \rangle dt$$

$$\leq \int_{E_T} \langle f'(x(t)), x'(t) \rangle dt \leq -\gamma \mathrm{mes}(E)$$

and
$$\text{mes}(E_T) \leq \gamma^{-1}(K_0 - \inf(f, C)).$$

This implies that
$$\text{mes}(\{t \in [0, \infty) : \Lambda(x(t)) \leq -\epsilon/2\}) \leq \gamma^{-1}(K_0 - \inf(f, C)). \qquad (4.118)$$

In order to complete the proof of the theorem, it is sufficient to show that for all large enough $t > 0$,
$$\Lambda(x(t)) \geq -\epsilon.$$

Assume the contrary. Then there exists a strictly increasing sequence $T_i > 0$, $i = 1, 2, \ldots$ such that
$$T_{i+1} \geq T_i + 2, \ \Lambda x(T_i)) \leq -\epsilon. \qquad (4.119)$$

By (4.82) and (4.115), for a. e. $t \geq 0$,
$$\|x'(t)\| \leq c_* + 1 \leq 2c_*. \qquad (4.120)$$

Let $i \geq 1$ be an integer and
$$t \in [T_i, T_i + (2c_*)^{-1}\delta]. \qquad (4.121)$$

Equations (4.120) and (4.121) imply that
$$\|x(t) - x(T_i)\| \leq \| \int_{T_i}^{t} x'(s)ds\| \leq \delta. \qquad (4.122)$$

It follows from (4.113) and (4.114) that
$$|\Lambda(x(t)) - \Lambda(x(T))| \leq \epsilon/4.$$

Together with (4.119), this implies that
$$\Lambda(x(t)) \leq \Lambda(x(T)) + \epsilon/4 \leq -\epsilon/2$$

and
$$\cup_{i=1}^{\infty}[T_i, T_i + (2c_*)^{-1}\delta] \subset \{t \in [0, \infty) : \Lambda(x(t)) \leq -\epsilon/2\}.$$

This contradicts (4.118). The contradiction we have reached proves Theorem 4.15.

4.12 Proof of Theorem 4.17

There exists $K_0 > 0$ such that
$$K_0 > \sup\{|f(x)| : x \in B(0, K) \cap C\}. \qquad (4.123)$$

Lemma 4.10 implies that there exist $\gamma, \Delta_0 \in (0, 1)$ such that the following property holds:

(a) For each $x \in B(0, K) \cap C$ satisfying $\Lambda(x) \leq -\epsilon$ and each $l \in E(x, \Delta_0, \Delta_0)$, the inequality $\langle f'(x(t)), l \rangle \leq -\gamma$ holds.

There exists $L > 0$ such that

$$\|f(z_1) - f(z_2)\| \leq L\|z_1 - z_2\|, \ z_1, z_2 \in B(0, K + 1) \cap U. \tag{4.124}$$

Choose a number $\Delta \in (0, \Delta_0)$ such that

$$(L + 2 + 2c_*)\Delta \leq \gamma\epsilon. \tag{4.125}$$

Assume that $T > 0$,

$$x : [0, T] \to C, \ x \in W^{1,1}(0, T), \tag{4.126}$$

$$\|x(t)\| \leq K, \ t \in [0, T], \tag{4.127}$$

$$x'(t) \in E(x(t), \Delta, \Delta), \ t \in [0, T] \text{ a. e.} \tag{4.128}$$

Set

$$\Omega = \{t \in [0, T] : \ \Lambda(x(t)) \leq -\epsilon\}. \tag{4.129}$$

Property (a) and (4.126)–(4.129) imply that for $t \in \Omega$ a. e.

$$\langle f'(x(t)), x'(t) \rangle \leq -\gamma. \tag{4.130}$$

Lemma 3.3 and (4.124) and (4.128) imply that for a. e. $t \in [0, T]$,

$$\langle f'(x(t)), x'(t) \rangle \leq (\|f'(x(t))\|_* + 2 + 2c_*)\Delta \leq (L + 2 + 2c_*)\Delta. \tag{4.131}$$

Lemma 4.3 and (4.123), (4.127), (4.130), and (4.131) imply that

$$-2K_0 \leq f(x(T)) - f(x(0)) = \int_0^T \langle f'(x(t)), x'(t) \rangle dt$$

$$= \int_{[0,T] \setminus \Omega} \langle f'(x(t)), x'(t) \rangle dt + \int_\Omega \langle f'(x(t)), x'(t) \rangle dt$$

$$\leq -\gamma \text{mes}(\Omega) + (L + 2 + 2c_*)\Delta(T - \text{mes}(\Omega)),$$

and

$$\gamma \text{mes}(\Omega) \leq 2K_0 + (L + 2 + 2c_*)\Delta T.$$

Together with (4.125), this implies that

$$\text{mes}(\Omega) \leq 2\gamma^{-1}K_0 + \gamma^{-1}(L + 2 + 2c_*)\Delta T$$

$$\leq 2K_0\gamma^{-1} + \epsilon T.$$

Theorem 4.17 is proved.

4.13 The Second Convergence Result with Estimations

Assume that $(X, \|\cdot\|)$ is a Banach space, $U \subset X$ is an open convex set,

$$C \subset U$$

is a nonempty closed and convex set, and a function $f : U \to R^1$ is Frechet differentiable at any point $x \in U$ and is Lipschitz on all bounded sets in U.

Let $K \geq 1$, $c_* > 1$ and $\lambda \in (0, 1]$ be fixed. We consider the function $\Lambda(x)$, $x \in C$ defined by (4.80) and the sets $E(x, \delta, \Delta)$ defined by (4.81)–(4.83) for $x \in C$ and $\delta, \Delta \in (0, 1]$.

In this section, we consider the minimization problem

$$f(x) \to \min, \; x \in C.$$

Assume that
$$\inf(f, C) = \inf(f, C \cap B(0, K)),$$

$$K_0 > \sup\{|f(x)| : \; x \in U \cap B(0, K)\}, \tag{4.132}$$

$$K_1 \geq \sup\{\|x\| : \; x \in U, \; |f(x)| \leq K_0 + 1\}, \tag{4.133}$$

$L_0 \geq 1$, $\rho \in (0, 1]$ and that

$$|f(z_1) - f(z_2)| \leq L_0 \|z_1 - z_2\|, \; z_1, z_2 \in B(0, K_1 + 1) \cap U, \tag{4.134}$$

$$\|f'(z_1) - f'(z_2)\|_* \leq L_0 \|z_1 - z_2\|^\rho, \; z_1, z_2 \in B(0, K_1 + 1) \cap U, \tag{4.135}$$

$$\epsilon \in (0, 1), \; \Delta = 8^{-1} \lambda \epsilon, \tag{4.136}$$

$$\delta = 8^{-1} \epsilon \lambda (2c_* + L_0)^{-1}, \tag{4.137}$$

$$\alpha = \min\{((32c_*^2 L_0)^{-1} \epsilon \lambda)^{1/\rho}, (2c_*)^{-1}\}. \tag{4.138}$$

Note that Lemma 3.18 holds and implies the following result.

Lemma 4.19. *Let* $x \in B(0, K_1) \cap C$ *satisfying*

$$\Lambda(x) \leq -\epsilon$$

and $l \in E(x, \delta, \Delta)$. *Then the inequality*

$$\langle f'(x), l \rangle \leq -2^{-1} \lambda \epsilon$$

holds.

Proof. Lemma 3.18 implies that for each $\beta \in (0, \alpha]$,

$$\beta^{-1}(f(x + \beta l) - f(x)) \leq 2^{-1}\lambda\epsilon.$$

This implies that

$$\langle f'(x), l \rangle \leq -2^{-1}\lambda\epsilon.$$

Lemma 4.19 is proved.

Theorem 4.20. *Assume that*

$$N_0 > 4K_0(\lambda\epsilon)^{-1},$$

$x \in W^{1,1}(0, N_0)$ *satisfies*

$$\|x(0)\| \leq K, \tag{4.139}$$

$$x(t) \in C, \ t \in [0, N_0],$$

and

$$x'(t) \in E(x(t), \delta, \Delta), \ t \in [0, N_0] \ a. \ e. \tag{4.140}$$

Then

$$\sup\{\Lambda(x(t)) : \ t \in [0, N_0]\} \geq -\epsilon.$$

Proof. Assume that the theorem does not hold. Then

$$\Lambda(x(t)) < -\epsilon, \ t \in [0, N_0]. \tag{4.141}$$

We show that

$$f(x(t)) \leq K_0, \ t \in [0, N_0]. \tag{4.142}$$

By (4.132) and (4.139),

$$f(x(0)) < K_0,$$

and there exists

$$t_0 \in (0, N_0]$$

such that

$$f(x(t)) < K_0, \ t \in [0, t_0]. \tag{4.143}$$

Assume that (4.142) does not hold. By (4.143), there exists

$$t_1 \in (t_0, T)$$

such that

$$f(x(t)) \leq K_0, \ t \in [0, t_1], \ f(x(t_1)) = K_0. \tag{4.144}$$

By (4.133) and (4.144),

$$\|x(t)\| \leq K_1, \ t \in [0, t_1]. \tag{4.145}$$

Lemma 4.19 and Eqs. (4.140), (4.141), and (4.145) imply that for a. e. $t \in [0, t_1]$,

$$\langle f'(x(t)), x'(t) \rangle \leq -2^{-1}\lambda\epsilon. \tag{4.146}$$

Lemma 4.3 and (4.143) and (4.146) imply that

$$f(x(t_1)) < f(x(0)) \leq K_0.$$

This contradicts (4.144). The contradiction we have reached proves that (4.142) is true. By (4.133) and (4.142),

$$\|x(t)\| \leq K_1, \ t \in [0, N_0]. \tag{4.147}$$

Lemma 4.19 and Eqs. (4.140), (4.141), and (4.147) imply that for a. e. $t \in [0, N_0]$,

$$\langle f'(x(t)), x'(t) \rangle dt \leq -2^{-1}\epsilon\lambda. \tag{4.148}$$

Lemma 4.3 and Eqs. (4.132), (4.142), and (4.148) imply that

$$-2K_0 \leq f(x(N_0)) - f(x(0)) = \int_0^{N_0} \langle f'(x(t)), x'(t) \rangle dt \leq -2^{-1}\epsilon\lambda N_0$$

and

$$N_0 \leq 4K_0(\lambda\epsilon)^{-1}.$$

This contradicts the choice of N_0. The contradiction we have reached completes the proof of Theorem 4.20.

References

1. Alber YI (1971) On minimization of smooth functional by gradient methods. USSR Comp Math Math Phys 11:752–758
2. Alber YI, Iusem AN, Solodov MV (1997) Minimization of nonsmooth convex functionals in Banach spaces. J Convex Anal 4:235–255
3. Alber YI, Iusem AN, Solodov MV (1998) On the projected subgradient method for nonsmooth convex optimization in a Hilbert space. Math Program 81:23–35
4. Barbu V, Precupanu T (2012) Convexity and optimization in Banach spaces. Springer, Heidelberg
5. Barty K, Roy J-S, Strugarek C (2007) Hilbert-valued perturbed subgradient algorithms. Math Oper Res 32:551–562
6. Bauschke H, Wang C, Wang X, Xu J (2015) On subgradient projectors. SIAM J Optim 25:1064–1082
7. Beck A, Teboulle M (2003) Mirror descent and nonlinear projected subgradient methods for convex optimization. Oper Res Lett 31:167–175
8. Brezis H (1973) Operateurs maximaux monotones. North Holland, Amsterdam
9. Burachik RS, Grana Drummond LM, Iusem AN, Svaiter BF (1995) Full convergence of the steepest descent method with inexact line searches. Optimization 32:137–146
10. Ceng LC, Hadjisavvas N, Wong NC (2010) Strong convergence theorem by a hybrid extragradient-like approximation method for variational inequalities and fixed point problems. J Glob Optim 46:635–646
11. Ceng LC, Wong NC, Yao JC (2015) Hybrid extragradient methods for finding minimum norm solutions of split feasibility problems. J Nonlinear and Convex Analysis 16:1965–1983
12. Censor Y, Gibali A, Reich S (2011) The subgradient extragradient method for solving variational inequalities in Hilbert space. J Optim Theory Appl 148:318–335

A. J. Zaslavski, *Optimization in Banach Spaces*, SpringerBriefs in Optimization, https://doi.org/10.1007/978-3-031-12644-4

13. Censor Y, Gibali A, Reich S, Sabach S (2012) Common solutions to variational inequalities. Set-Valued Var Anal 20:229–247

14. Censor Y, Davidi R, Herman GT, Schulte RW, Tetruashvili L (2014) Projected subgradient minimization versus superiorization. J Optim Theory Appl 160:730–747

15. Chadli O, Konnov IV, Yao JC (2004) Descent methods for equilibrium problems in a Banach space. Comput Math Appl 48:609–616

16. Demyanov VF, Rubinov AM (1967) The minimization of a smooth convex functional on a convex set. SIAM J Control 5:280–294

17. Demyanov VF, Vasilyev LV (1985) Nondifferentiable optimization. Optimization Software, New York

18. DeVore RA, Temlyakov VN (2016) Convex optimization on Banach spaces. Found Comput Math 16:369–394

19. Gao Z, Petrova G (2019) Rescaled pure greedy algorithm for convex optimization. Calcolo: 56:14 pp

20. Gibali A, Jadamba B, Khan AA, Raciti F, Winkler B (2016) Gradient and extragradient methods for the elasticity imaging inverse problem using an equation error formulation: a comparative numerical study. Nonlinear Analysis and Optimization Contemp Math 659:65–89

21. Griva I (2018) Convergence analysis of augmented Lagrangian-fast projected gradient method for convex quadratic problems. Pure Appl Funct Anal 3:417–428

22. Hiriart-Urruty J-B, Lemarechal C (1993) Convex analysis and minimization algorithms. Springer, Berlin

23. Konnov IV (2003) On convergence properties of a subgradient method. Optim Methods Softw 18:53–62

24. Konnov IV (2009) A descent method with inexact linear search for mixed variational inequalities. Russian Math (Iz VUZ) 53:29–35

25. Konnov IV (2018) Simplified versions of the conditional gradient method. Optimization 67:2275–2290

26. Korpelevich GM (1976) The extragradient method for finding saddle points and other problems. Ekon Matem Metody 12:747–756

27. Li X, Yong J (1995) Optimal control theory for infinite dimensional systems. Birkhauser, Boston

28. Liu L, Qin X, Yao J-C (2019) A hybrid descent method for solving a convex constrained optimization problem with applications. Math Methods Appl Sci 42:7367–7380

29. Mordukhovich BS (2006) Variational analysis and generalized differentiation, I: I: Basic Theory. Springer, Berlin

30. Mordukhovich BS, Nam NM (2014) An easy path to convex analysis and applications. Morgan and Clayton Publishers, San Rafael

31. Nadezhkina N, Takahashi Wataru (2004) Modified extragradient method for solving variational inequalities in real Hilbert spaces. In: Nonlinear analysis and convex analysis, pp 359–366. Yokohama Publication, Yokohama

32. Nedic A, Ozdaglar A (2009) Subgradient methods for saddle-point problems. J Optim Theory Appl 142:205–228

33. ODHara JG, Pillay P, Xu HK (2006) Iterative approaches to convex feasibility problems in Banach spaces. Nonlinear Anal 64:2022–2042

34. Pallaschke D, Recht P (1985) On the steepest–descent method for a class of quasidifferentiable optimization problems. In: Nondifferentiable optimization: motivations and applications (Sopron, 1984), pp 252–263. Lecture Notes in Economics and Mathematical Systems, vol 255. Springer, Berlin

35. Polyak BT (1987) Introduction to optimization. Optimization Software, New York

36. Polyak RA (2015) Projected gradient method for non-negative least squares. Contemp Math 636:167–179

37. Qin X, Cho SY, Kang SM (2011) An extragradient-type method for generalized equilibrium problems involving strictly pseudocontractive mappings. J Global Optim 49:679–693

38. Reich S, Zaslavski AJ (2014) Genericity in nonlinear analysis. Developments in Mathematics. Springer, New York

39. Shor NZ (1985) Minimization methods for non-differentiable functions. Springer, Berlin

40. Solodov MV, Zavriev SK (1998) Error stability properties of generalized gradient-type algorithms. J Optim Theory Appl 98:663–680

41. Su M, Xu H-K (2010) Remarks on the gradient-projection algorithm. J Nonlinear Anal Optim 1:35–43

42. Temlyakov VN (2015) Greedy approximation in convex optimization. Constr Approx 41:269–296

43. Thuy LQ, Wen C-F, Yao J-C, Hai TN (2018) An extragradient-like parallel method for pseudomonotone equilibrium problems and semigroup of nonexpansive mappings. Miskolc Math Notes 19:1185–1201

44. Wang H, Xu H-K (2018) A note on the accelerated proximal gradient method for nonconvex optimization. Carpathian J Math 34:449–457

45. Xu H-K (2011) Averaged mappings and the gradient-projection algorithm. J Optim Theory Appl 150:360–378

46. Xu H-K (2017) Bounded perturbation resilience and superiorization techniques for the projected scaled gradient method. Inverse Prob 33:19 pp

47. Yao Y, Postolache M, Yao J-C (2019) Convergence of an extragradient algorithm for fixed point and variational inequality problems. J Nonlinear Convex Anal 20:2623–2631

48. Zaslavski AJ (2020) The projected subgradient algorithm in convex optimization. Springer, Berlin. Briefs in Optimization (2020)

49. Zaslavski AJ (2010) The projected subgradient method for nonsmooth convex optimization in the presence of computational errors. Numer Funct Anal Optim 31:616–633

50. Zaslavski AJ (2012) The extragradient method for convex optimization in the presence of computational errors. Numer Funct Anal Optim 33:1399–1412
51. Zaslavski AJ (2012) The extragradient method for solving variational inequalities in the presence of computational errors. J Optim Theory Appl 153:602–618
52. Zaslavski AJ (2013) The extragradient method for finding a common solution of a finite family of variational inequalities and a finite family of fixed point problems in the presence of computational errors. J Math Anal Appl 400:651–663
53. Zaslavski AJ (2016) Numerical optimization with computational errors. Springer, Cham
54. Zaslavski AJ (2020) Convex optimization with computational errors. In: Springer Optimization and Its Applications. Springer, Cham
55. Zeng LC, Yao JC (2006) Strong convergence theorem by an extragradient method for fixed point problems and variational inequality problems. Taiwanese J Math 10:1293–1303

Printed in the United States
by Baker & Taylor Publisher Services